Welding

Level Four

SIXTH EDITION

Pearson

NCCER National Center for Construction Education and Research

NCCER
President and Chief Executive Officer: Boyd Worsham
Vice President of Innovation and Advancement: Jennifer Wilkerson
Chief Learning Officer: Lisa Strite
Senior Curriculum Leader: Chris Wilson
Senior Manager, Production: Graham Hack
Production Operations Manager: Kelly Beck
Project Manager: Lauren Henehan
Project Coordinator: Emmah Walker
Technical Writing Manager: Gary Ferguson
Technical Writer: Simone Flynn
Art Manager: Bree Rodriguez
Technical Illustrators: Judd Ivines, Liza Wailes
Production Artist: Chris Kersten
Permissions Specialist: Sherry Davis
Managing Editor: Hannah Murray
Lead Editor: Alexandria Willbond
Editors: Zi Meng, Lauren Tygrest, Terry ONeil, Melina Redondo
Desktop Publishing Manager: Eric Caraballoso III
Desktop Publishing Coordinator: Daphney Milian
Desktop Publishing Specialist: Julie Watkins
XML Developer: James Singer III
Digital Content Coordinator and Translation Manager: Yesenia Tejas
Digital Content Coordinator: Briana Rosa

Pearson
Commercial Product Manager – Associations: Andrew Dunaway
Senior Digital Content Producer: Shannon Stanton
Associate Project Manager: Monica Perez-Kim
Senior Content Producers: Alexandrina B. Wolf, Alma Dabral
Content Producers: Ashwathy Chandrasekar, Arup Kumar Ghosh, Mary Roth
Executive Marketing Manager: Mark Marsden
Designer: Mary Siener
Rights and Permissions: Kaitlyn Coddington

Composition: NCCER
Content Technologies: Gnostyx
Printer/Binder: Lakeside Book Company
Cover Printer: Lakeside Book Company
Text Typefaces: Palatino LT Pro and Helvetica Neue LT Pro

10 9 8 7 6 5 4 3 2 1 2 2025

Pearson

NCCER National Center for Construction Education and Research

Paperback
ISBN-13: 978-0-13-821304-6

PREFACE

To the Trainee

Welding is a cutting-edge profession that provides skilled welders with the opportunity to work in settings that span across the globe. Welders, when equipped with the proper training and passion for the trade, can find steady jobs ranging from ship building to manufacturing automobiles to working in steel mills and industrial plants.

If you have the desire and motivation to further your education in the welding industry, you will find that you have many options for employment and great potential for career growth. According to the Department of Labor, "Skilled welders with up-to-date training should have good job opportunities." Welders will continue to be in high demand since there is a continuous need to rebuild aging infrastructure such as highways and buildings. In addition, the creation of new power generation facilities will yield new job prospects for trained and skilled welders.

Welding techniques are continually evolving. For example, new methods are being developed to bond dissimilar materials and nonmetallic materials, such as plastics, composites, and new alloys. Also, advances in laser beam and electron beam welding, new fluxes, and other new technologies all point to an increasing need for educated, skilled welders.

Each level of the Welding curriculum presents an apprentice approach and will help you be knowledgeable, safe, and effective on the job. This curriculum has been revised by industry subject matter experts from across the nation who incorporated the latest methods and technology of the trade.

New with *Welding Level Four*

This revised sixth edition of *Welding Level Four* has been reformatted to provide a better experience to both trainees and instructors. *Welding Level Four* contains code updates referencing new editions of the American Welding Society's School Excelling through National Skills Education (SENSE) and additional safe practices, design improvements, and content updates to several technologies. This edition of *Welding Level Four* updated the title of NCCER Module 29405 to *Soldering and Brazing of Copper*. All the content has updated technology, safety information, and best practices.

We wish you success as you progress through this training program. If you have any comments on how NCCER might improve upon this textbook, please complete the User Update form using the QR code on this page. NCCER appreciates and welcomes its customers' feedback. You may submit yours by emailing **support@nccer.org**. When doing so, please identify feedback on this title by listing *#WeldingL4* in the subject line.

Our website, **www.nccer.org**, has information on the latest product releases and training.

NCCER

SCAN ME

NCCER Standardized Curricula

NCCER is a not-for-profit 501(c)(3) education foundation established in 1996 by the world's largest and most progressive construction companies and national construction associations. It was founded to address the severe workforce shortage facing the industry and to develop a standardized training process and curricula. Today, NCCER is supported by hundreds of leading construction and maintenance companies, manufacturers, and national associations. The NCCER Standardized Curricula was developed by NCCER in partnership with Pearson, the world's largest educational publisher.

Some features of the NCCER Standardized Curricula are as follows:

- An industry-proven record of success
- Curricula developed by the industry, for the industry
- National standardization providing portability of learned job skills and educational credits
- Compliance with the Office of Apprenticeship requirements for related classroom training (CFR 29:29)
- Well-illustrated, up-to-date, and practical information

NCCER maintains a secure online database that provides certificates, digital badges, transcripts, and wallet cards to individuals who successfully complete programs under an NCCER-accredited organization or through one of NCCER's self-paced, online programs. This system also allows individuals and employers to track and verify industry-recognized credentials and certifications in real time.

For information on NCCER's credentials, contact NCCER Customer Service at 1-888-622-3720 or visit **https://www.nccer.org**.

Digital Credentials

Show off your industry-recognized credentials online with NCCER's digital credentials!

NCCER is now providing online credentials. Transform your knowledge, skills, and achievements into digital credentials that you can share across social media platforms, send to your network, and add to your resume. For more information, visit **www.nccer.org**.

DESIGN FEATURES

Content is organized and presented in a functional structure that allows you to access the information where you need it.

You can navigate Welding using color coded tabs on the upper right hand corner.

Source: Photo courtesy of The Lincoln Electric Company, Cleveland, OH, U.S.A.

Objectives

Successful completion of this module prepares you to do the following:

1. Identify GTAW-related safety practices and describe the electrical characteristics that affect GTAW.
   a. Identify GTAW-related safety practices.
   b. Describe the electrical characteristics that affect GTAW.
2. Identify and describe GTAW equipment and consumables.
   a. Identify and describe GTAW welding machines.
   b. Identify and describe GTAW torches.
   c. Identify and describe GTAW torch nozzles and electrodes.
   d. Identify and describe GTAW shielding gases.

Performance Tasks

Under supervision, you should be able to do the following:

1. Select a GTAW shielding gas.
2. Select a GTAW filler metal.
3. Connect the shielding gas and set the flow rate.

Overview

To produce consistent, high-quality welds, welders must select, install, and use the equipment and materials involved in the gas tungsten arc welding (GTAW) process properly. This module provides an overview of the equipment and safety concerns that are associated with GTAW. It describes GTAW equipment and consumables, including power sources, torches, nozzles, electrodes, shielding gases, and filler metals. It also explains how to select, prepare, and set up GTAW equipment safely and efficiently.

NCCER **Industry-Recognized Credentials**

If you are training through an NCCER-accredited sponsor, you may be eligible for credentials from NCCER. The ID number for this module is 29207. Note that this module may have been used in other NCCER curricula and may apply to other level completions. Contact NCCER at 1.888.622.3720 or go to **www.nccer.org** for more information.

You can also show off your industry-recognized credentials online with NCCER's digital badges. Transform your knowledge, skills, and achievements into badges that you can share across social media platforms, send to your network, and add to your resume. For more information, visit **www.nccer.org**.

Digital Resources for Welding

Scan this code using the camera on your phone or mobile device to view the digital resources related to this craft.

SCAN ME

The Objectives list the skills and knowledge you need in order to complete the module successfully.

The Performance Tasks give you an opportunity to apply your knowledge to real-world tasks.

Section Openers provide a visual organizational structure for the information. Objectives and Performance Tasks are broken out for each section.

1.0.0 Common Hand Tools

Performance Task

1. Inspect and demonstrate the safe and proper use of the following hand tools:
   - Hammers
   - Demolition tools
   - Chisels and punches
   - Screwdrivers
   - Adjustable wrenches
   - Non-adjustable wrenches
   - Sockets
   - Pliers

Objective

Name common hand tools and state how to use them.

a. Identify various hammers and demolition tools and explain how to use them.
b. Describe chisels and punches and how they are used.
c. Match screwdrivers to the appropriate hardware.
d. Differentiate between non-adjustable, adjustable, and socket wrenches.
e. Describe various types of pliers and explain how they are used.

Construction craftworkers require a combination of knowledge and practical skills to be successful. While there are many things that can be done with hands alone, every craft must use hand tools to get the job done. How craftworkers handle and care for their tools often demonstrates their level of skill and respect for the craft.

Many unique hand tools are used across the crafts—far too many to present

1.1.0 Sine Wave Terminology

Several AC voltage terms are specific to sine waves. It is important to understand this terminology when working with AC. The following sections discuss some of these terms, including **frequency**, period, wavelength, peak value, average value, and effective value.

1.1.1 Frequency

The frequency of a waveform is the number of times per second an identical pattern repeats itself. Each time a waveform changes from zero to a peak value and back to zero it is called an *alternation*. Two alternations form one cycle. The number of cycles per second is the frequency. The unit of frequency is **hertz (Hz)**. One Hz equals one cycle per second (cps).

Frequency: The number of cycles an alternating electric current, sound wave, or vibrating object undergoes per second.

Hertz (Hz): A unit of frequency; one hertz (Hz) equals one cycle per second (cps).

Trade Terms appear on the page adjacent to the text where they are first presented.

Step-by-step math equations help make the concepts clear and easy to grasp.

2.0.0 Unknown Values in AC Circuits 25

Solve for X_L as follows:

Step 1	Compute the value of X_L.	$X_L = 2\pi fL$ $X_L = 6.28 \times 100\text{ Hz} \times 4\text{H}$ $X_L = 2{,}512\Omega$
Step 2	Draw vectors R and X_L as shown in *Figure 19*. R is drawn horizontally because the circuit current and voltage across vector R are in phase. It therefore becomes the reference line from which other angles are measured. X_L is drawn upward 90° from R because the voltage across X_L leads the circuit current through R.	
Step 3	Compute the value of circuit impedance Z, which is equal to the vector sum of X_L and R.	$\tan = \frac{X_L}{R} = \frac{2{,}512}{1{,}500} = 1.67$ $\arctan(1.67) = 59.1°$ $\cos(59.1°) = 0.5135$
	Find Z using the cosine function.	$\cos = \frac{R}{Z}$ $Z = \frac{R}{\cos}$ $Z = \frac{1{,}500}{0.5135} = 2{,}921\Omega$
Step 4	Compute the circuit current using Ohm's law for AC circuits.	$I = \frac{E}{Z} = \frac{100\text{V}}{2{,}921\Omega} = 0.034\text{A}$

QR codes link directly to videos that highlight current content.

SUCCESS STORIES: *Credentials Matter: Start Building Your Career Today*

Scan this code using the camera on your phone or mobile device to view this video.

Important information is highlighted, illustrated, and presented to facilitate learning.

Placement of images near the text description and details such as callouts and labels help you absorb information.

Preparing Drills with Keyless Chucks

Most cordless drills use a keyless chuck. While the steps for preparing a cordless drill are similar, there are some small differences. Follow the steps below when preparing to use drills with keyless chucks:

Step 1 Disconnect the drill from its power source by removing the battery pack before loading a bit.
Step 2 As shown in (*Figure 7A*), open the chuck by turning it counterclockwise until the jaws are wide enough to insert the bit shank.
Step 3 Insert the bit shank into the chuck opening (*Figure 7B*). Keeping the bit centered in the opening, turn the chuck by hand until the jaws grip the bit shank.
Step 4 Tighten the chuck securely with your hand so that the bit does not move (*Figure 7C*). You are now ready to use the cordless drill.

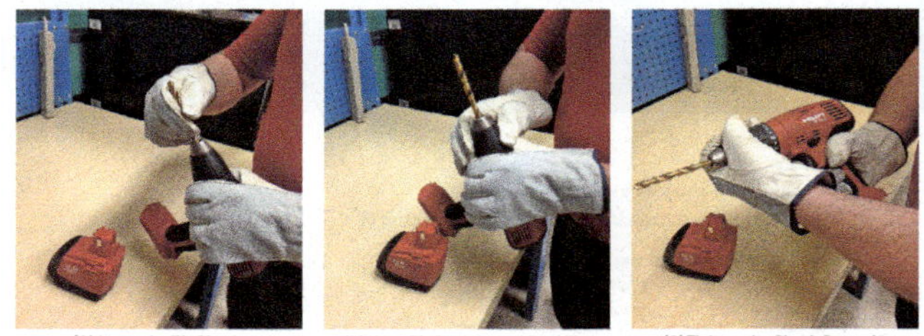

(A) Insert the Bit Shank. **(B) Keep Bit Straight and Partially Tighten the Chuck.** **(C) Tighten the Chuck Securely.**

Figure 7 Loading the bit on a keyless chuck.
Source: Cianbro Corporation

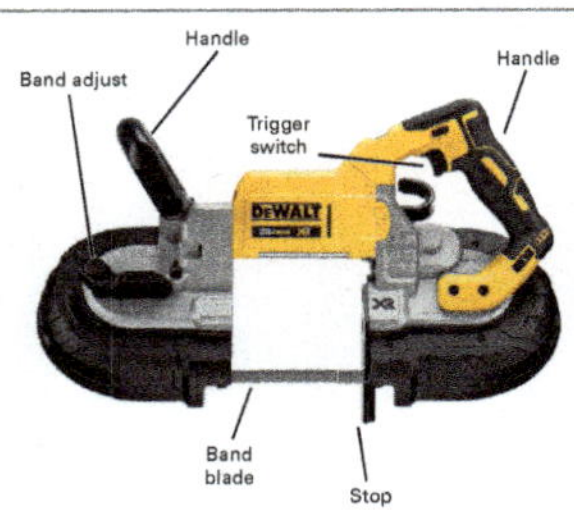

Figure 27 Portable handheld band saw.
Source: Image property of Stanley Black & Decker. Used with permission

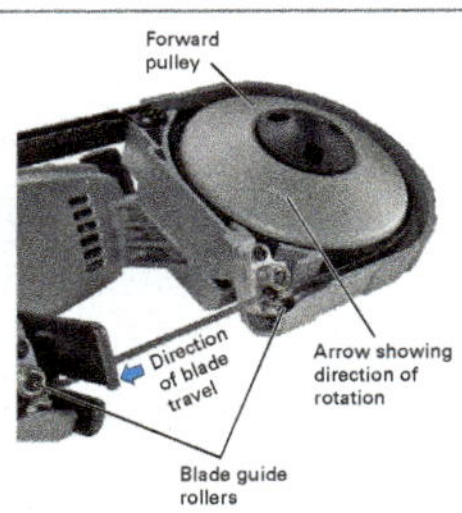

Figure 28 Band saw pulley and blade guide rollers.
Source: Courtesy of Milwaukee Electric Tool Corporation

blade length is determined by referencing the manufacturer's documentation. Like most blades, its coarseness is rated in teeth per centimeter or teeth per inch (TPI). As a general rule, higher TPIs produce smoother cuts. Thicker materials require coarser blades. However, if the blade is too coarse for the material, the individual teeth may begin to break off. ***Figure 28* shows how the blade is routed** around the pulleys and through the blade guides.

While some band saws have multiple speeds, most do not. The portable band saw generally cuts best at a low speed since higher speeds cause the blade's teeth to rub rather than cut. This can create heat through friction, which reduces the life of the blade.

New boxes highlight safety and other important information. Warning boxes stress potentially dangerous situations, while Caution boxes alert to dangers that may cause damage to equipment. Note boxes provide additional information on a topic.

WARNING!

A portable band saw always cuts in the direction of the user. For that reason, workers must be especially careful to avoid injury when using this type of saw. Always wear appropriate PPE and stay focused on the work.

CAUTION

Never assume anything. It never hurts to ask questions, but disaster can result if you don't ask. For example, do not assume that an electrical power source is turned off. First ask whether the power is turned off, then check it yourself to be completely safe.

NOTE

This training alone does not provide any level of certification in the use of fall arrest or fall restraint equipment. Trainees should not assume that the knowledge gained in this module is sufficient to certify them to use fall arrest equipment in the field.

Did You Know?

Lightning Rods

An interesting fact about grounding is that a lightning rod (air terminal) isn't meant to bring a bolt of lightning to ground. To do this, its conductors would have to be several feet (1 m or more) in diameter. The purpose of the rod is to dissipate the negative static charge that would cause the positive lightning charge to strike the house.

These boxed features provide additional information that enhances the text.

Orthographic Drawings

Orthographic drawings are used for elevation drawings. They show straight-on views of the different sides of an object with dimensions that are proportional to the actual physical dimensions. In orthographic drawings, the designer draws lines that are scaled-down representations of real dimensions. Every 12 inches, for example, may be represented by ¼ inch on the drawing. Similarly, in an example using metric measurements with a ratio of 1:2, every 30 millimeters may be represented by 15 millimeters on the drawing.

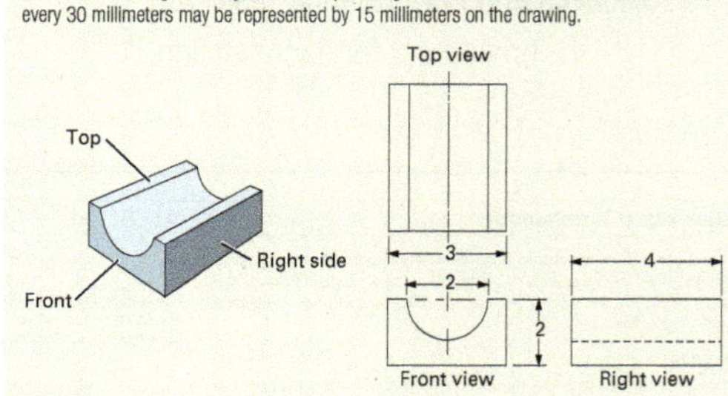

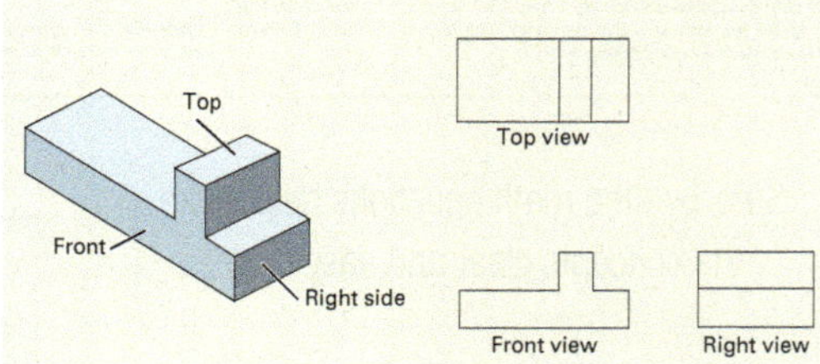

Around the World

Predicting Weather

Barometers measure atmospheric air pressure. Generally, an increased pressure indicates that the weather is pleasant. A reduced pressure indicates that the weather is cloudy and/or rainy.

Changes in barometric pressure are used to predict changes in weather conditions. For example, if it is raining outside, but the barometer is rising, it indicates that better weather is on the way. In the midst of a hurricane, the barometric pressure is an indicator of intensity. Extremely low pressures exist in powerful hurricanes. While traditional barometers work very well, sophisticated weather stations capture far more data and can transmit the information wirelessly.

Going Green

Residential Solar

As residential customers look for ways to reduce reliance on fossil fuels like coal and oil, they are turning to solar energy. Usually, they want to install photovoltaic (PV) panels on the roofs of their homes, or in their yards. A residential PV system usually includes the PV panels, a mounting structure that can support the panels, an inverter to convert the direct current (DC) electricity generated by solar photovoltaic modules into alternating current (AC) electricity, and a battery (or batteries) for storing the energy. Most states require a licensed electrician for solar energy system installation, but you should check the local requirements. (Source: **https://www.energy.gov/eere/solar/how-does-solar-work**)

Going Green looks at ways to preserve the environment, save energy, and make good choices regarding the health of the planet.

15302 Precision Measuring Tools 43

Cornerstone of Craftsmanship

Kim T. Panepinto
Mechanical Planner
Industrial Parts Specialties, LLC

How did you choose a career in the industry?
I was working at a gas station and wanted to make more money. So, I applied for a welder helper position, and then when they asked about a millwright helper, I asked, "What's that?" True story! I had no idea what it was. Alan Landry, the plant manager at the local chemical plant, got me into the millwright craft.

What types of training have you been through?
I've been through ABC Millwright training, Total Quality Management (TQM) programs, Rand McNally mechanical seals training, Lubritech lubrication system training, and laser alignment classes. There are many others that I've attended over the years to both broaden and sharpen my skills.

How important is education and training in construction?
It's very important! Nobody is born knowing how to do all this stuff. Knowledge can't be taken away from someone, so always take any training offered to you. The only other way to learn is hands-on, but classrooms and curriculums set the standard, and that's how you know you're learning the right way. Hands-on training and practice then become an extension of the classroom, instead of the only learning opportunity.

How important are NCCER credentials to your career?
My early training came before NCCER curricula was available, but I am an NCCER-certified instructor. Right now, many plants in my state are hiring rotating equipment people and many require the NCCER training to be eligible to be hired. I think that's a good indicator of how important the credentials have become.

How has training/construction impacted your life?
Without training, I certainly wouldn't be as successful as I have been. It is essential for career growth.

What kinds of work have you done in your career?
I've done just about everything from being a simple oiler to a project manager and served in many possible positions in between those two.

Tell us about your current job.
I'm currently a mechanical planner, and my primary task is to develop the scope for specific jobs. These are detailed, step-by-step plans of what we are going to do with a piece of equipment—take it down and/or put it back together—and I actually happen to have a scope on my desk right now. They outline what is required, what parts need to be ordered and acquired in advance, and the labor hours required.

Cornerstone of Craftsmanship sections feature career stories from people working in related fields.

Review questions at the end of each section and module allow you to measure your progress.

Section Review questions can be found at the end of each section to test your knowledge of the content.

Review Questions at the end of each module are provided to reinforce the knowledge you have gained.

2.0.0 Section Review

1. A straight pull contains two raceways. One of the raceways has a trade size of 3" and one has a trade size of 2". The length of the box *must* be _____.
   a. 16"
   b. 24"
   c. 26"
   d. 32"

2. Where angle or U pulls are made, to determine the distance between raceway entries enclosing the same conductor, you need to multiply the trade size of the largest raceway by _____.
   a. two
   b. four
   c. six
   d. eight

3. If possible, pull boxes should be installed _____.
   a. high on the wall for security
   b. at a height/location that makes it easy to pull conductors
   c. behind wall coverings for a neater look
   d. with extra knockouts left open for air flow

Module 26207-23 Review Questions

1. In which *NEC*® article is cable tray installation primarily addressed?
   a. *NEC Article 300*
   b. *NEC Article 392*
   c. *NEC Article 517*
   d. *NEC Article 550*

2. Which type of tray is *best* suited for corrosive areas and atmospheres?
   a. Stainless steel cable tray
   b. Coated aluminum cable tray
   c. Nonmetallic cable tray
   d. Basket tray

3. When a cable tray has a solid bottom, it is referred to as _____.
   a. basket tray
   b. ladder tray
   c. trough
   d. raceway

4. Cable tray is generally manufactured in _____.
   a. 2' and 4' lengths
   b. 6' and 8' lengths
   c. 6' and 12' lengths
   d. 12' and 24' lengths

Additional information can be found in Appendixes at the back of the book.

371

APPENDIX 29202A

Example of a Three-View Detail Drawing

Figure A01 is an example of an object that may be encountered in welding. *Figure A02* is an example of a three-view detail drawing representing that same object.

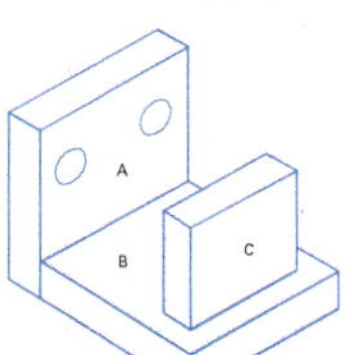

Details			
Part	Thickness	Length	Width
A	¾"	3½" (Tall)	3½"
B	¾"	4" (Long)	3½"
C	¾"	2½" (Tall)	2½"

Hole Dimensions and Location
- Hole diameter is ¾"
- Hole centers are 1¼" from the top edge of block "A"
- Hole centers are ¾" from the outside edges of block "A"

Figure A01

Some modules have corresponding Appendixes that provide supplementary information or activities to enhance your understanding of the material.

NCCERCONNECT

This interactive online course is a unique web-based supplement that provides a range of visual, auditory, and interactive elements to enhance training. Also included is a full eText.

Visit **www.nccerconnect.com** for more information!

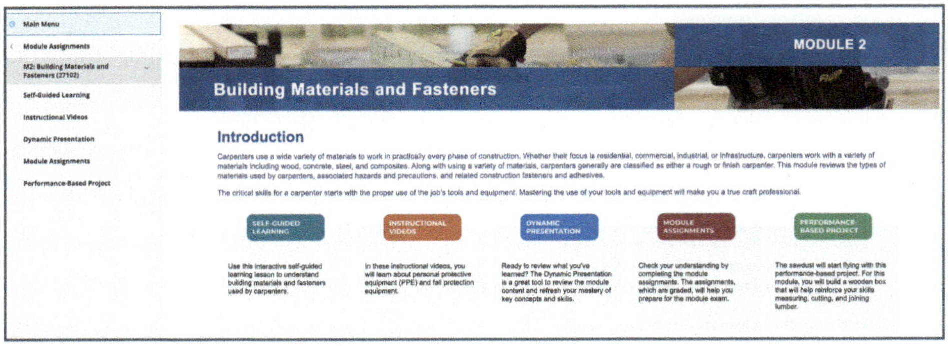

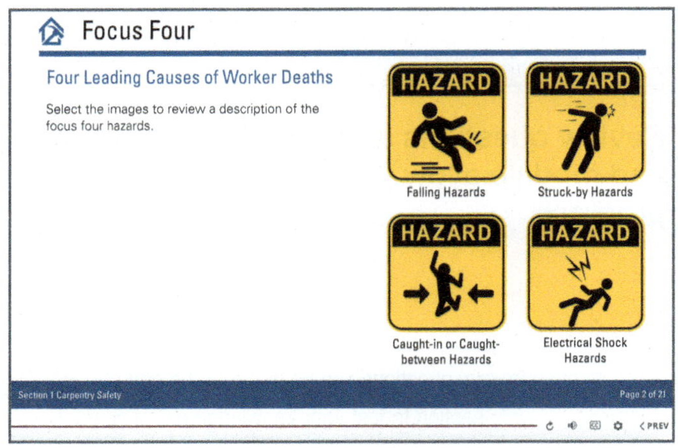

Use the interactive self-guided learning lesson to understand key concepts and terms needed for a career in construction.

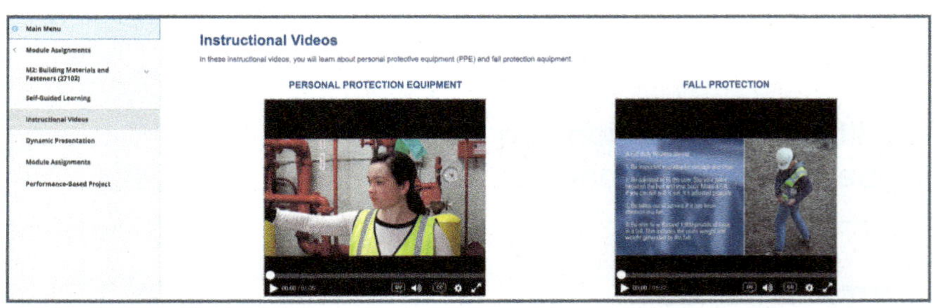

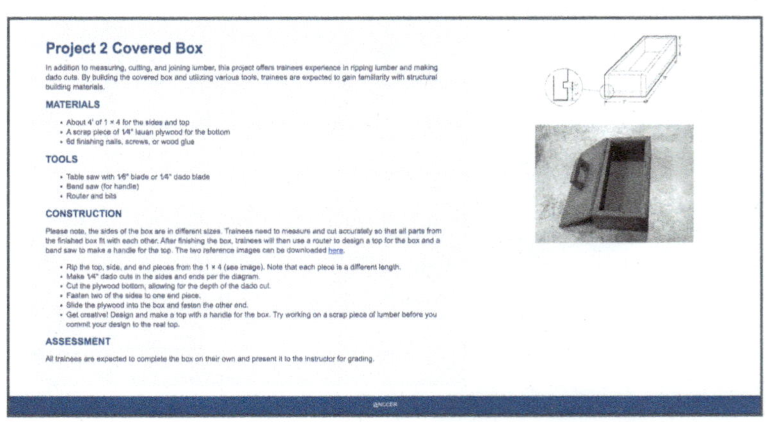

ACKNOWLEDGMENTS

This curriculum was revised as a result of the vision and leadership of the following sponsors:

All Things Metal
ArcLabs Welding School
Bechtel Global Corporation
College of Southern Nevada, Department of Applied Technologies
Elite Welding Academy
Kentucky Welding Institute
Lee College
Lincoln Technical Institute
Mesa Community College
Miller Electric
Northeast Community College
NW Florida State College
Raycap
Southern Crescent Technical College
United Group Services
Windham School District
Yates Elite Welding Academy

This curriculum would not exist were it not for the dedication and unselfish energy of those volunteers who served on the Authoring Team. A sincere thanks is extended to the following:

Timothy Allen
Ashley Applegate
Matthew Aschoff
Curtis Casey
Jim Casey
Sam Chance
Brian Dennis
John Elliott
John Gaffin
Don Herron
Donald "Greg" MacLiver
Jimmy Perry
Nick Peterson
Ben Pletcher
Mike Rice
Rich Samanich
Scottie Smith
Jeremy Wright

A final note: This book is the result of a collaborative effort involving the production, editorial, and development staff at Pearson Education, Inc., and NCCER. Thanks to all of the dedicated people involved in the many stages of this project.

NCCER PARTNERS

To see a full list of NCCER Partners, please visit:

www.nccer.org/about-us/partners.

SCAN ME

You can also scan this code using the camera on your phone or mobile device to view these partnering organizations.

CONTENTS

MODULE 29401 GMAW – Aluminum Plate

MODULE 29404 GMAW – Aluminum Pipe

MODULE 29402 GTAW – Aluminum Plate

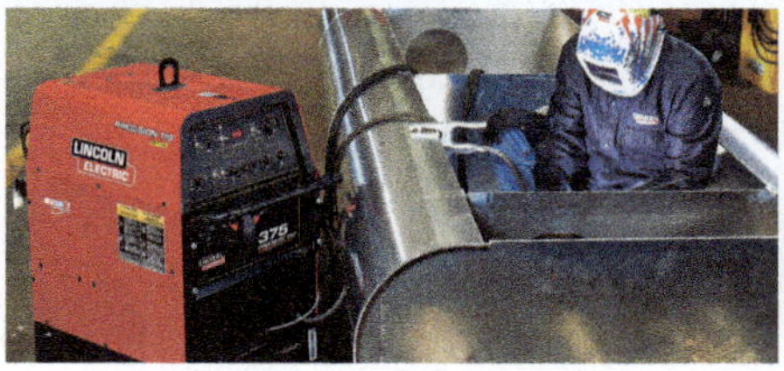

MODULE 29403 GTAW – Aluminum Pipe

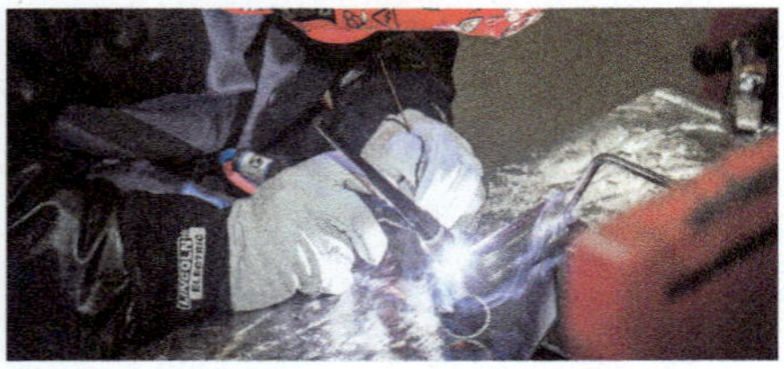

MODULE 29405 Soldering and Brazing of Copper

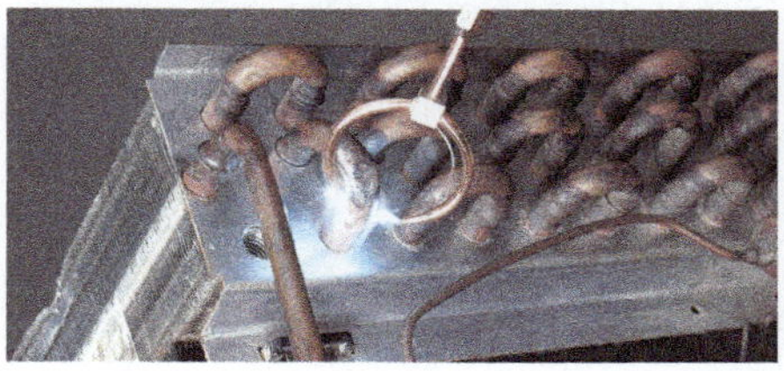

Welding Appendixes

MODULE 29401

GMAW – Aluminum Plate

Source: Photo courtesy of The Lincoln Electric Company, Cleveland, OH, U.S.A.

Objectives

Successful completion of this module prepares you to do the following:

1. Describe GMAW safety precautions and welding preparation activities.
    a. Identify common welding safety precautions.
    b. Describe GMAW equipment for aluminum welding.
    c. Explain how to prepare the area and materials for GMAW.
2. Describe GMAW concepts and how to produce fillet and V-groove welds.
    a. Identify the types of aluminum alloys.
    b. Identify key challenges related to aluminum welding.
    c. Describe weld beads, terminations, and restarts.
    d. Describe how to make aluminum fillet welds in various positions.
    e. Describe how to make aluminum plate V-groove welds in various positions.

Performance Tasks

Under supervision, you should be able to do the following:

1. Complete the following GMAW aluminum plate welding tasks:
    - Stringer beads
    - Weave beads
    - Weld terminations
    - Weld restarts
    - Overlapping beads
2. Complete GMAW fillet welds on aluminum plate in the 1F, 2F, 3F, and 4F positions.
3. Complete GMAW V-groove welds on aluminum plate with backing in the 1G, 2G, 3G, and 4G positions.

Overview

Gas metal arc welding (GMAW) is one of the more common welding processes used today. As with most welding systems, GMAW equipment is available in many different sizes and varieties. However, the basic operating principles of GMAW apply to all makes and models of equipment. This module describes how GMAW equipment is used to weld aluminum plate and examines different techniques used to produce fillet and V-groove plate welds.

Digital Resources for Welding

SCAN ME

Scan this code using the camera on your phone or mobile device to view the digital resources related to this craft.

NCCER Industry-Recognized Credentials

If you are training through an NCCER-accredited sponsor, you may be eligible for credentials from NCCER. The ID number for this module is 29401. Note that this module may have been used in other NCCER curricula and may apply to other level completions. Contact NCCER at 1.888.622.3720 or go to **www.nccer.org** for more information.

You can also show off your industry-recognized credentials online with NCCER's digital credentials. Transform your knowledge, skills, and achievements into credentials that you can share across social media platforms, send to your network, and add to your resume. For more information, visit **www.nccer.org**.

1.0.0 GMAW Hazards and Welding Preparation

Performance Tasks

There are no Performance Tasks in this section.

Objective

Describe GMAW safety precautions and welding preparation activities.

a. Identify common welding safety precautions.
b. Describe GMAW equipment for aluminum welding.
c. Explain how to prepare the area and materials for GMAW.

Gas metal arc welding (GMAW) is an arc welding process that uses a continuous, consumable, solid wire electrode for the filler metal and a shielding gas to protect the weld zone. The GMAW process is commonly used to make welds on aluminum, carbon steel, low-alloy steel, and stainless steel.

This module reviews GMAW equipment safety and setup and explains how to prepare aluminum plate for both fillet welds and V-groove welds in common positions.

1.1.0 Safety Practices

Safety considerations are a critical part of preparing for a welding operation. The following is a summary of safety procedures and practices that must be observed while cutting or welding. Before continuing this module, it is strongly recommended that you complete NCCER Module 29101, *Welding Safety*, which provides a thorough review of welding hazards and precautions. Above all, be sure to wear appropriate personal protective equipment (PPE) at all times when welding or cutting.

WARNING!

Welding aluminum is associated with immediate and long-term hazards and health risks that can be reduced and controlled, if not eliminated, by following safety practices. Short-term exposure to welding fumes may cause eye and throat irritation, dizziness, and nausea. Prolonged exposure to welding fumes may cause ulcers, damage to the lungs, kidneys, and nervous systems, as well as various types of cancer, including lung, larynx, and urinary tract.

Due to health risks associated with the welding of aluminum, the Occupational Safety and Health Administration (OSHA) has established permissible exposure limits (PELs) for aluminum as well as iron or mild steel. OSHA's PEL limit of 5 mg/m^3 represents the total concentration of fume as an exposure limit over an eight-hour time weighted average. Welders adhere to PELs in addition to following safety practices and protocols of protective clothing, fire prevention, fall prevention, and venting.

1.1.1 Protective Clothing and Equipment

Protective clothing and equipment are a necessary part of every welding task. The following general guidelines review typical PPE required for welding:

- *Welding hood* — A welding hood equipped with a properly shaded lens is best for all forms of welding. The welding hood lens must be an appropriate shade for the application.
- *Clothing and gloves* — Wear proper leather or flame-retardant clothing along with welding gloves for protection from flying sparks, molten metal, and heat.
- *Footwear* — Wear high-top safety shoes or boots. Make sure the tongue and lace area of the footwear is covered by a pant leg. If the tongue and lace area is exposed or if the footwear must be protected from burn marks, wear spats under the pant leg or leather chaps over the front and top of the footwear.
- *Caps and hard hats* — Wear a 100% cotton cap without any mesh material. If a hard hat is required for the environment, use one that allows the attachment of rear deflector material and a face shield. A hard hat with a rear deflector may be required when working overhead and may be required by some employers and jobsites.
- *Ear protection* — Wear earmuffs, or at least earplugs, to prevent sparks from entering the ear canals.

Welders must also familiarize themselves with any unique requirements an employer or job may have regarding PPE.

WARNING!

Welding aluminum is associated with specific hazards and risks that require awareness and safety protocols. Reflected ultraviolet (UV) light is especially challenging when welding aluminum. Aluminum's highly reflective surface poses a big concern. When welding aluminum, ensure all exposed skin is covered by wearing proper PPE and using welding screens to protect others. Additionally, it is important to follow all other OSHA safety precautions.

Lightweight Welding Gloves

Soft, flexible leather gloves are used for light-duty work and for operations involving GMAW, GTAW, brazing, and soldering where more flexibility and dexterity are required. These welding processes produce less heat and spatter than others. More dexterity is required for the welding techniques and for operating torch controls. A wide variety of styles, using various materials, are available. The gloves pictured are used for GTAW and are not as heat-resistant as some other gloves suitable for GMAW. Note, when using welding gloves, always be aware that contamination on gloves can potentially contaminate base metal or filler metal.

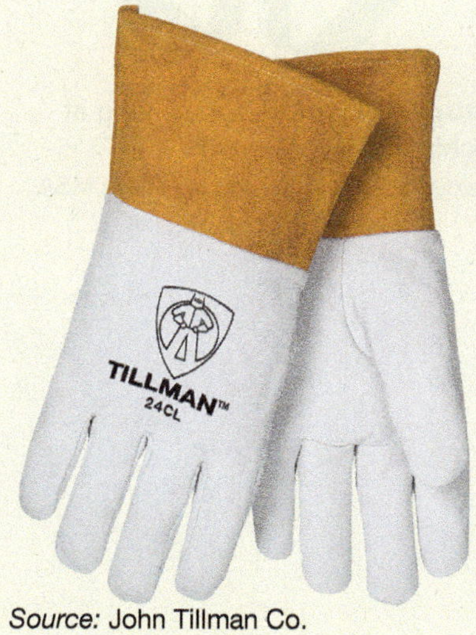

Source: John Tillman Co.

1.1.2 Fall Protection and Housekeeping

Welding in the construction industry occurs in a variety of different environments. Welding inside large, open shops exposes a welder to equipment hazards on the floor and overhead. Welding outdoors exposes the welder and the welding job to potential weather hazards. In both environments, welders often must crawl under or reach over objects to complete their work.

At times, welders working indoors and outdoors must work at height. Fall protection equipment is required by OSHA when working at a height of 6' or higher (*Figure 1*).

Figure 1 Harness for working at height.
Source: Used with permission from MSA

Fall protection equipment saves lives. Always keep fall protection devices untangled and properly attached. Combined with the other equipment a welder must carry, there is a great deal to manage, but fall protection equipment must always be the priority. Never attach any kind of welding gear to the harness that could interfere with its ability to perform its intended function.

Regardless of where you work, always keep the work area as neat as possible. This makes the area safer and helps reduce distractions. Mindful actions, such as making sure welding cables are not tripping hazards, contribute to a safer environment. Electrode stubs and unused electrodes must not be left on the floor. They can cause someone to lose their footing and fall. Parts waiting to be welded or left over from previous work are a tripping hazard and should be kept out of the walking area. If the parts have sharp edges, they could cut welding cable insulation or torch gas hoses.

1.1.3 Fire/Explosion Prevention

Flammable gases are commonly found around welding and cutting operations, as are sparks and other significant sources of ignition. For that reason, and because many fire-related incidents over the years have been traced back to welding operations, it is crucial that fire and explosion prevention be given consistent attention. Follow these guidelines to maintain a safe environment:

- Never carry matches, gas-filled lighters, or battery powered devices such as cell phones or vape pens. Sparks can cause the matches or lighters to ignite, or the battery packs to overheat and explode. Any ignition or explosion can cause serious personal injury.
- Never use oxygen to blow dust or dirt from clothing. Traces of the oxygen can remain trapped in the fabric for some time. If a spark hits the fabric during this time, the clothing can burn rapidly and violently out of control.
- Make sure any flammable material in the work area is removed or shielded by a fire-resistant covering. Approved fire extinguishers must be available before attempting any heating, welding, or cutting operations. If a hot work permit and a fire watch are required, be sure these are in place before beginning.
- Ensure all site safety requirements are being observed.
- Never release a large amount of oxygen or use oxygen in place of compressed air. Its presence around flammable materials or sparks can cause rapid and uncontrolled combustion. Keep oxygen away from oil, grease, and other petroleum products.
- Never release a large amount of fuel gas, especially acetylene. Heavier than air, propane tends to concentrate in and along low areas and can ignite at a considerable distance from the release point. Acetylene is lighter than air but is even more dangerous than propane. When mixed with air or oxygen, acetylene will ignite at much lower concentrations than any other fuel gas.

- Even though shielding gases are not toxic, any gas released in a sufficient quantity can displace the oxygen in a small room or enclosed work area, creating an atmosphere with little or no oxygen. Avoid large releases of any type of gas, regardless of its characteristics.
- To prevent fires, maintain a neat and clean work area, and make sure that any metal scrap or slag is cold before disposal.
- Before cutting containers such as tanks or barrels, check to see if they have contained any explosive, hazardous, or flammable materials, including petroleum products, citrus products, or chemicals that decompose into toxic fumes when heated. As a standard practice, always clean and then fill any tanks or barrels with water or purge them with a flow of inert gas, such as nitrogen or argon, to displace any oxygen. Containers must be cleaned by steam cleaning, flushing with water, or washing with detergent until all traces of the material have been removed. Proper procedures for cutting or welding hazardous containers are described in the *American Welding Society (AWS) F4.1, Safe Practices for the Preparation of Containers and Piping for Welding, Cutting, and Allied Processes.*

WARNING!

Welding or cutting must never be performed on drums, barrels, tanks, vessels, or other containers until they have been emptied and cleaned thoroughly, filled with water, or purged with inert gas. All flammable materials and all substances (such as detergents, solvents, greases, tars, or acids) that might produce flammable, toxic, or explosive vapors when heated need to be removed or stable to create a safe atmosphere for welding. Clean containers only in well-ventilated areas. Vapors can accumulate during cleaning, causing explosions and serious personal injury.

1.1.4 Work Area Ventilation

Many gases are introduced to the atmosphere through welding processes. Some of these gases are harmful, but all of these gases can displace or combine with atmospheric oxygen in a confined space. In the case of aluminum, GMAW applies high heat to melt the aluminum and its alloys which facilitates oxidation. Aluminum oxide particles are formed, which mix with the welding fumes. The welding fumes produce white visible smoke that contain known lung irritants, magnesium oxide, and aluminum oxide.

Follow these guidelines, as well as those of your employer, for work area ventilation to prevent personal injuries including long-term respiratory harm:

- Follow all confined space procedures before welding or cutting in a confined space. Use powered fume extraction systems, as needed.
- Never use oxygen to provide ventilation in confined spaces.
- Always perform cutting or welding operations in a well-ventilated area.

The welding of aluminum is associated with specific gases and fumes including ozone, nitrous oxides, carbon monoxide, and carbon dioxide. For example, ozone is a known lung irritant, so proper ventilation needs to be utilized to minimize the risk. Ozone is produced during GMAW of aluminum. The UV rays from the welding arc interact with oxygen in the area forming ozone gas.

The composition and quantity of the gases and fumes produced from welding aluminum are dependent upon the welding process, filler wire used, coatings on the metal being welded (i.e. paint, anodizing or Alodine coatings), number of welders, volume of the work area, quality and the amount of ventilation, position of the welder with respect to the fume plume, and presence of contaminants in the atmosphere such as chlorinated hydrocarbon vapors from cleaning and degreasing activities.

For long-term cutting or welding of aluminum, and other materials producing toxic gases and fumes, always wear an approved full-face, supplied-air respirator that uses breathing air supplied from outside the work area. For occasional short-term exposure, a HEPA-rated or metal fume filter may be used on a standard respirator.

In addition to controlling work area ventilation for aluminum-specific risks, powered fume extraction equipment must also be used when welding materials that contain zinc, cadmium, hexavalent chromium, manganese, and similar materials that produce hazardous fumes if the area is not naturally well-ventilated. *Figure 2* shows examples of both mobile and portable units.

WARNING!

Breathing in fumes from welding aluminum can cause metal fume fever, also called *welder's flu*. Symptoms include chest tightness, cough, metallic taste, fever, and headache. Proper ventilation on the worksite and individual PPE are necessary to limit exposure and potential risk from fumes created by welding aluminum.

Figure 2 Powered air purifying respirator (PAPR) and localized fume extraction equipment.

Source: Photo courtesy of The Lincoln Electric Company, Cleveland, OH, U.S.A.

1.2.0 GMAW Welding Equipment

Identify a proper welding machine for GMAW. Follow these steps to set the machine up for use:

Step 1 Verify that the welding machine can be used for GMAW.

Step 2 Know the location of the primary disconnect.

Step 3 Configure the welding machine for GMAW (*Figure 3*) as directed by your instructor. Configure the gun polarity, or *direct current electrode positive (DCEP)*. Connect and load the wire feeder with the appropriate filler metal as required. Equip the gun with the correct nozzle for the application and the proper liner material and contact tube for the diameter of wire being used. The contact tube should be designed specifically for aluminum wire to ensure the size matches properly. Push-pull guns are most commonly used. Spool guns are another option. Both types are shown in *Figure 4*.

CAUTION

Using contact tips designed for steel wire is not acceptable due to the expansion rate of the aluminum wire; binding will likely occur as the weld progresses. Use only contact tips designed for aluminum.

NOTE

The wire feeder drive rollers should be of the U-groove style for aluminum wire, unlike the V-groove rollers used for carbon steel wire.

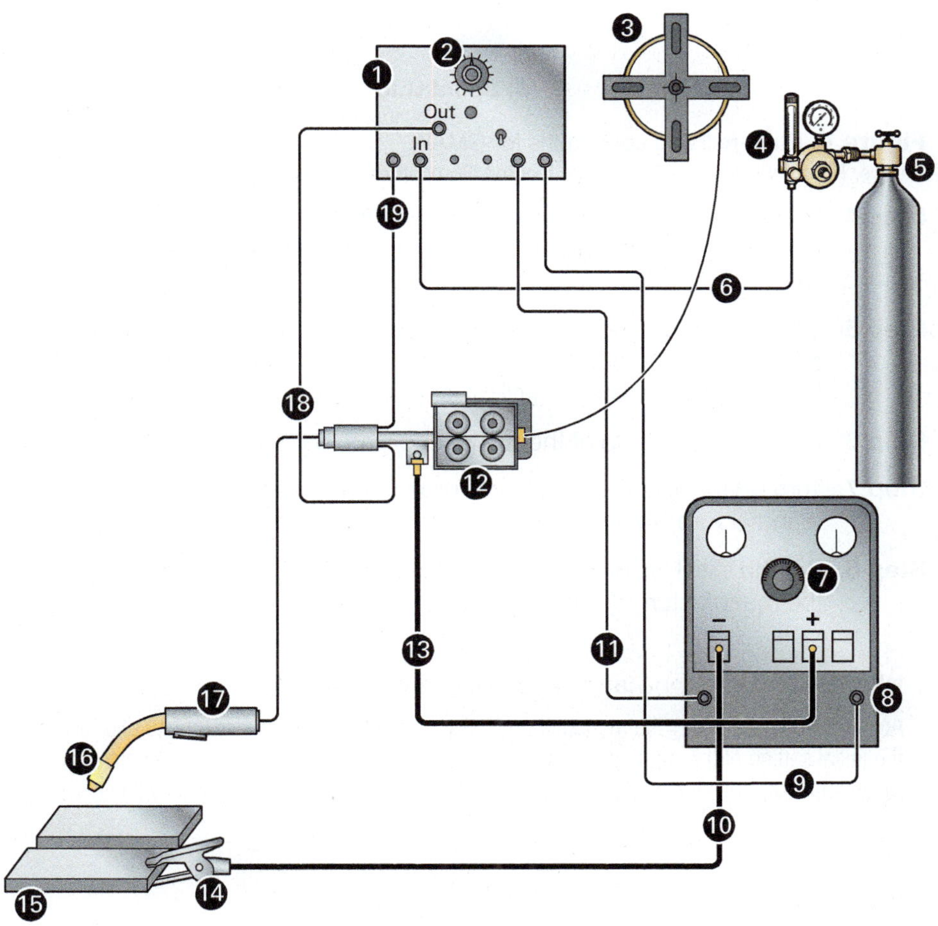

Note: The polarity of the gun and workpiece leads is determined by the type of filler metal and application.

1. Control Unit
2. Wire Feed Speed Control
3. Electrode Wire Reel
4. Regulator/Flowmeter
5. Shielding Gas Source
6. Gas-In Hose
7. Voltage Control
8. Power Source
9. Power to Control Unit
10. Workpiece Lead
11. Contactor Control Cable
12. Wire Feed Drive Assembly
13. Electrode Lead
14. Workpiece Clamp
15. Workpiece
16. Gas Nozzle
17. GMAW Gun
18. Gas-Out Hose
19. Gun Control Cable

Figure 3 Configuration diagram of a typical GMAW machine.

(A) Push-Pull Gun

(B) Spool Gun

Figure 4 Push-pull and spool guns for GMAW.
Source: Photo courtesy of The Lincoln Electric Company, Cleveland, OH, U.S.A.

CAUTION

To prevent contaminants on the electrode wire from contaminating the welds, always keep the wire clean and dry when it is in storage or in the machine.

Step 4 In accordance with the manufacturer's instructions, configure and load the wire feeder and gun with the proper type and diameter of electrode wire.

Step 5 Connect the proper shielding gas (argon or argon/helium) for the application as specified by the wire electrode manufacturer, Welding Procedure Specification (WPS), site quality standards, or your instructor.

Step 6 Connect the clamp of the workpiece lead to the workpiece.

Step 7 Turn on the welding machine. Purge the gun as directed by the gun manufacturer's instructions.

Step 8 Set the initial welding voltage and wire feed speed as recommended by the manufacturer for the type and size of electrode wire being used.

Wire Electrode Manufacturer's Recommendations

Always obtain and follow the manufacturer's recommendations for shielding gas and for the initial setting of the welding voltage and wire feed speed parameters. These balanced parameters are critical and are based on the welding position, size, and composition of the solid wire.

1.2.1 Welding Voltage

Arc length is determined by the voltage, which is set at the welding machine. Arc length is the distance from the wire electrode tip to the base metal or to the molten pool at the base metal. If the voltage is set too high, the arc will be too long. This can cause the wire to melt and fuse to the contact tube. Voltage that is too high can also cause porosity and excessive spatter. Voltage must be increased or decreased as the wire feed speed is increased or decreased. Set the voltage to maintain consistent spray transfer; a minimum arc voltage is needed to maintain spray transfer. However, penetration is not directly related to voltage. Penetration will increase as voltage is increased for a time, but penetration will actually decrease if the voltage is increased above its optimum setting.

1.2.2 Welding Amperage

When using a standard constant-voltage welding machine, the electrode feed speed controls the welding amperage, after the initial recommended setting. The welding machine provides the amperage needed to melt the wire electrode while maintaining the selected welding voltage. Within limits, when the wire electrode feed speed is increased, the welding amperage and deposition rate are also increased. This results in higher welding heat, deeper penetration, and higher beads. When the wire electrode feed speed is decreased, the welding amperage automatically decreases. With lower welding amperage and less heat, the deposition rate drops, and the weld beads are smaller with less penetration.

Note that some constant-voltage welding machine used for GMAW provide varying degrees of current modification. Examples include slope and induction adjustments. Welding machines with a slope adjustment let the welder vary the amount of amperage change in relation to the unit's voltage range. Standard constant-voltage GMAW units have a current slope fixed by the manufacturer for general welding applications and conditions. Pulse-transfer units allow the peak current for the pulse and the background current between pulses to be adjusted to match specific welding applications and/or wire electrode requirements.

1.2.3 Electrode Extension, Stickout, and Standoff Distance

Electrode extension is the length of the wire that extends beyond the tip of the welding gun's contact tip. For high-conductivity metal wires, the preheating effect of wire resistance is minimal. Wire speed and voltage settings have a more direct effect on weld penetration and deposition rate.

Either spray transfer or pulsed spray transfer can be used for aluminum welding. GMAW electrode extension for spray transfer varies from ½" to 1" (12.7 mm to 25.4 mm). *Figure 5* shows the typical electrode extension for the spray transfer GMAW gun configuration. It also shows various gun terms and components. Stickout is the distance from the gas nozzle or insulating nozzle to the end of the electrode. Standoff distance is the distance from the gas nozzle or insulating nozzle to the workpiece.

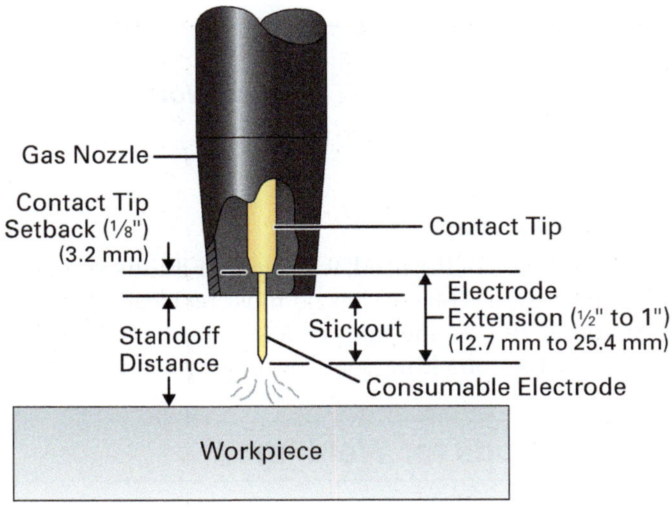

Figure 5 GMAW gun configured for spray transfer.

Contact tip setback is important in the spray or pulse transfer mode of GMAW. The setback protects the contact tip from the additional heat created during the spray or pulse transfer mode. Reducing the heat helps to keep the wire from melting in the contact tip.

1.2.4 Weld Travel Speed

The travel speed must be considered before beginning a weld. Weld travel speed is the speed at which the electrode tip passes across the base metal in the direction of the weld. It is measured in inches per minute (IPM). The metric values for travel speed are millimeters or centimeters per minute. Travel speed has a great effect on penetration and bead size. Slower travel speeds build higher beads with deeper penetration. Faster travel speeds build smaller beads with less penetration. Ideally, the welding parameters should be adjusted so that the electrode tip is positioned at the leading edge of the weld puddle during travel.

> **Travel Speed and Wire Feed Speed**
>
> New welders are tempted to turn down the wire feed speed if they have difficulty controlling the weld puddle. Wire electrodes must be run at certain balanced parameters that cannot be changed individually. Voltage, current (if variable), wire feed speed, and travel speed are adjusted and balanced together to control the weld puddle.

1.2.5 Gun Position

The gun position in relation to the direction of the weld is defined for aluminum as shown in *Figure 6*. A push angle of 5° to 10° is used to enhance base metal cleaning ahead of the weld puddle.

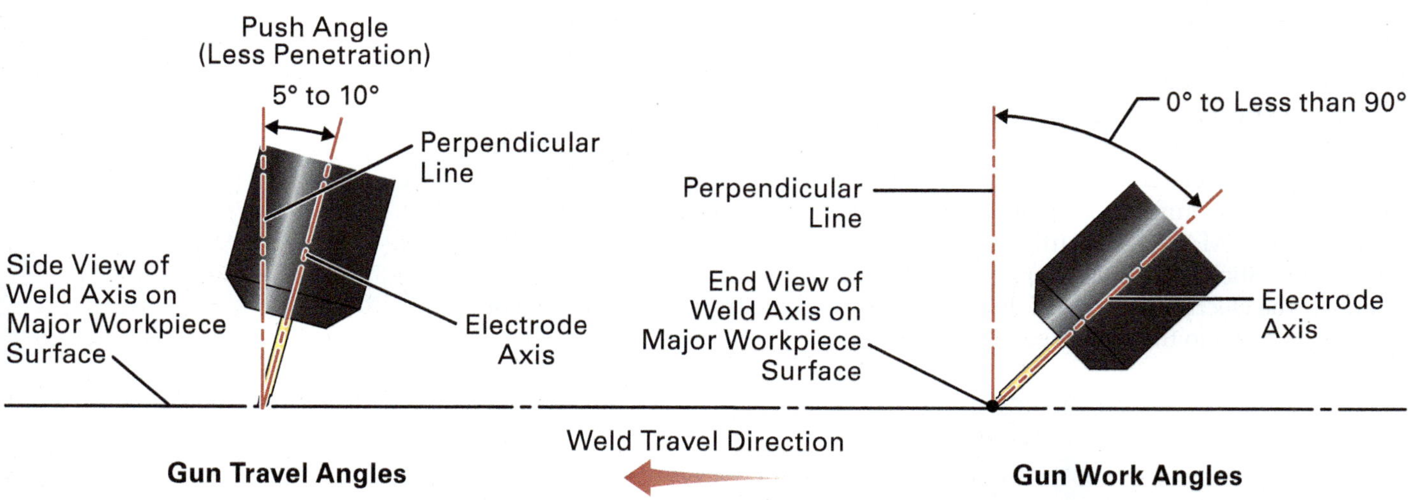

Figure 6 Gun travel and work angles.

1.2.6 Gas Nozzle Cleaning

As the welding machine is being used, weld spatter builds up on the gas nozzle, diffuser, and contact tube. The gas nozzle must be cleaned from time to time with a reamer, round file, or the tang of a standard file. If it is not properly cleaned, it will restrict the shielding gas flow and cause porosity in the weld.

1.3.0 Preparations for Welding

Before welding can begin, the area must be set up, welding equipment must be organized, and the metal to be welded must be prepared. The following sections explain how to prepare for welding.

To practice welding, you will need a welding table, bench, or stand. The welding surface can be steel, but an aluminum surface is preferred. Provisions must be available for placing weld coupons out of position.

To set up the area for welding, follow these steps:

Step 1 Make sure that the area is properly ventilated. Make use of doors, windows, and fans.

Step 2 Check the area for fire hazards. Remove any flammable materials before proceeding.

Step 3 Locate the nearest fire extinguisher. Do not proceed unless the extinguisher is charged and you know how to use it.

Step 4 Set up welding screens around the welding area.

1.3.1 Practice Welding Coupons

Because aluminum is easily contaminated, proper coupon preparation is a priority. Aluminum coupons must be cut to the desired lengths, cleaned, and beveled to the proper angles before they are positioned for welding. An aluminum backing must also be cut and cleaned. The basic shapes can be cut by shearing or sawing.

The importance of cleaning aluminum before welding cannot be overstated. There are two related facts about aluminum that help explain why. First, aluminum does not give the welder a visual indication that it is nearing its melting point through color change, as many other common metals do. With no indication that the melting point is approaching, it is easy to apply too much heat. Second, aluminum oxide—the "skin" that forms on the outside—has a melting point roughly three times higher than the aluminum itself. Excessive weld contamination and poor weld quality result when the aluminum base metal is not thoroughly cleaned.

The preferred technique for beveling and/or adding a root face is mechanical machining. Grinding is also acceptable, although machining the grooves and root faces of aluminum is the cleanest and most uniform way to prepare the material for GMAW. Thermal cutting techniques are not considered acceptable. If thermal cutting techniques must be used to cut aluminum alloys, at least 1/8" (3.2 mm) of the cut surface must be removed by machining or grinding before welding.

CAUTION

Grinding wheels, files, and similar tools used on aluminum should not be used on any other type of metal or on surfaces contaminated with grease or oil.

WARNING!

Grinding wheels used on aluminum must be made for aluminum only. They should not be used on any other type of metal or on surfaces contaminated with grease or oil. Using other types of grinding wheels on aluminum may cause the grinding wheels to shatter.

Practice coupons are best cut from aluminum plate that is 1/4" (6.4 mm) thick. If this size is not readily available, aluminum plate up to 3/8" (9.6 mm) thick can be used. Material up to 1/2" (12.7 mm) thick can be used with the joint design described herein if welding equipment of sufficient current capacity is available. Material thicknesses greater than 1/2" (12.7 mm) can be used, but a different joint design is required, and welding equipment of extreme current capacity is required when argon gas is used alone. Although it is expensive, helium can be added to the mixture to significantly reduce the welding current required.

Clean oil and grease from the coupons and the metal backing with an approved chemical such as acetone. After oil and grease have been removed, use a stainless steel wire brush to remove all other contaminants and oxidation. Use only light pressure on the brush.

NOTE

For welding thin aluminum materials (less than 1/4" or 6.4 mm), there are many other acceptable joint designs that do not require a backing. The joint design used in this module, however, is better suited for practice on the 1/4" (6.4 mm) and thicker material specified above.

WARNING!

To prevent injury when using cleaning solvents, always comply with the safety guidelines found on the Safety Data Sheet (SDS) for the solvents being used.

The coupons must be prepared to allow the following welds:

- *Stringer beads, weave beads,* and *overlapping beads* — The plate coupons (simple pads) can be any size or shape that can be easily handled.
- *Fillet welds* — Cut the metal into 4" by 7" (102 mm by 178 mm) rectangles for the base and 3" by 7" (76 mm by 178 mm) rectangles for the web (*Figure 7*).
- *V-groove welds with backing* — A properly prepared aluminum V-groove weld joint with backing is illustrated in *Figure 8*. Note that this joint design is not recommended for material thicker than ½" (12.7 mm). As shown in *Figure 9*, cut two pieces of metal into 3" by 7" (76 mm by 178 mm) rectangles with one (or both) of the 7" (178 mm) lengths beveled at 30°. Grind a $^1/_{32}$" to $^1/_8$" (0.8 mm to 3.2 mm) root face on the bevel as directed by your instructor. Note that a thicker root face makes it more difficult to achieve proper penetration and consistent fusion. Cut backing strips 1" by 8" (25.4 mm by 203 mm) long from the same alloy as the beveled pieces (*Figure 10*). The backing should be sufficiently thick to prevent melt-through.

Backing Strip Length

When the backing strip is ½" to 1" (12.7 mm to 25.4 mm) longer than the plates at each end, it allows you to start and stop the bead outside of the weld groove.

Going Green

Conserve Aluminum Practice Coupons

Aluminum is expensive and difficult to obtain for practice purposes. To conserve resources and reduce waste, completely use all weld coupons until all surfaces have been welded upon. Cutting coupons apart and reusing the pieces also conserves materials. Use flat material that cannot be cut into weld coupons to practice running beads. Note that stringer beads can be practiced while leaving enough room between them to practice weave beads as well. Any aluminum material that is no longer suitable for welding practice should be recycled.

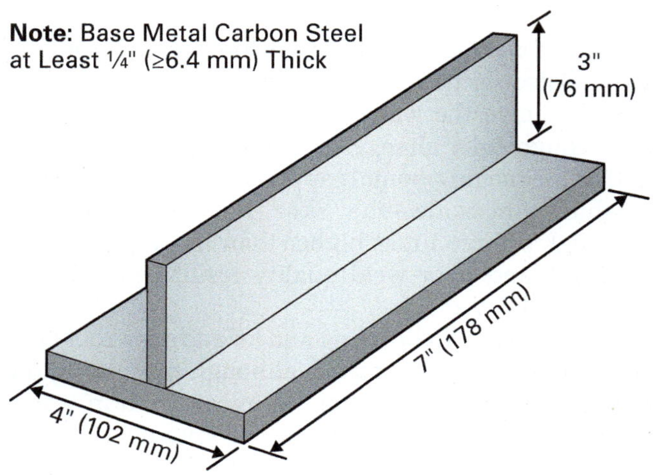

Figure 7 Fillet weld coupons.

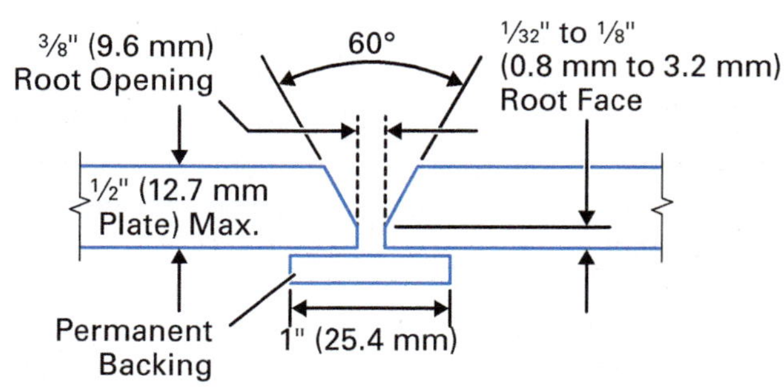

Fit Tolerances	
Root Opening	+¼"; −1⁄16" (+6.4 mm; −1.6 mm)
Groove Angle	+10°; −5°

Figure 8 GMAW aluminum weld joint preparation.

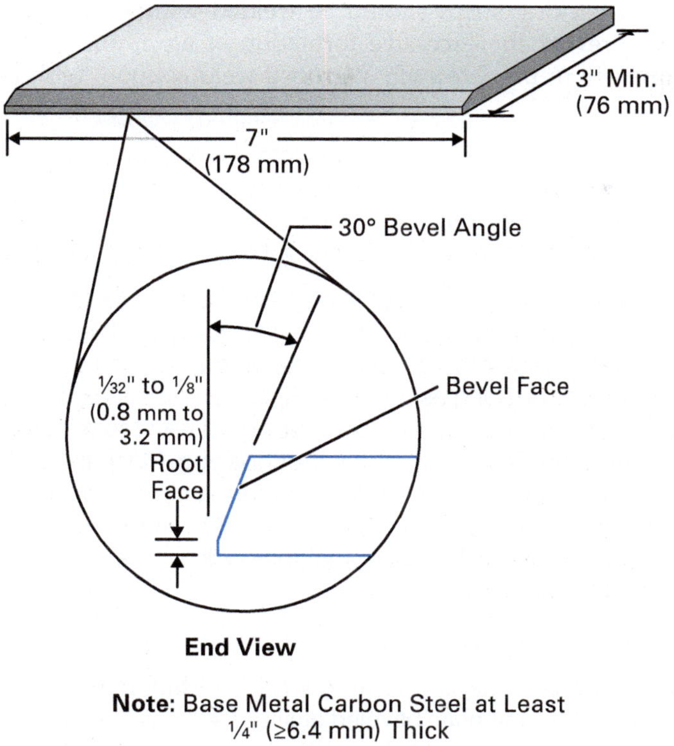

Figure 9 Preparing the V-groove weld coupon.

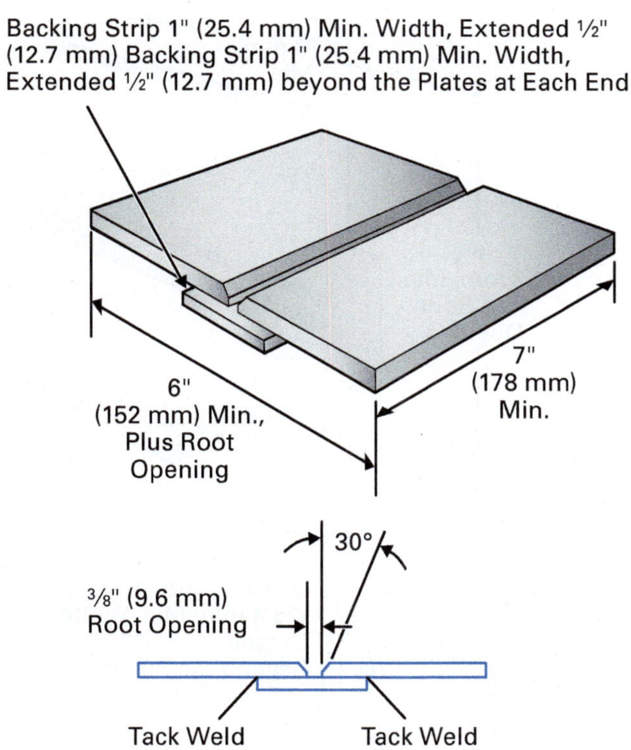

Figure 10 V-groove weld coupon with metal backing.

Coupons and backing strips should be welded within 30 minutes of their preparation to prevent the excessive formation of aluminum oxide. Observe the following steps to prepare each V-groove weld coupon, with backing, for welding:

Step 1 Cut and bevel the coupon components (if required) as shown using any of the acceptable methods discussed in this section. If the bevel has been cut thermally, remember to machine-cut the bevel back roughly ⅛" (3.2 mm) before machining the root face to its proper dimension.

Step 2 Make sure that the surface of the backing strip under the root opening has been thoroughly cleaned of all oxide coating in the same manner as the coupon. Center the beveled strips on the backing strip with a ⅜" (9.6 mm) root opening as shown in *Figure 10*, and tack-weld them in place. If desired, use a piece of metal or rod with the required thickness as a temporary spacer to set the root opening. Place the tack welds on the reverse side of the joint in the lap formed by the backing strip and the beveled plate. Use three to four ½" (12.7 mm) long tack welds on each beveled plate. Be certain that the backing strip is tight against the beveled plates.

1.3.2 Alternate Joint Preparation

An alternate joint preparation can be used when welding in the horizontal position. This variation has one plate beveled at about 45° and the other at about 15°. The plate beveled at 45° is placed above the 15° plate with a ⅜" (9.6 mm) root opening (*Figure 11*). Note that the degrees shown are approximate.

NOTE

Check with your instructor about whether to use the standard V-groove preparation or the alternate preparation for horizontal weld coupons.

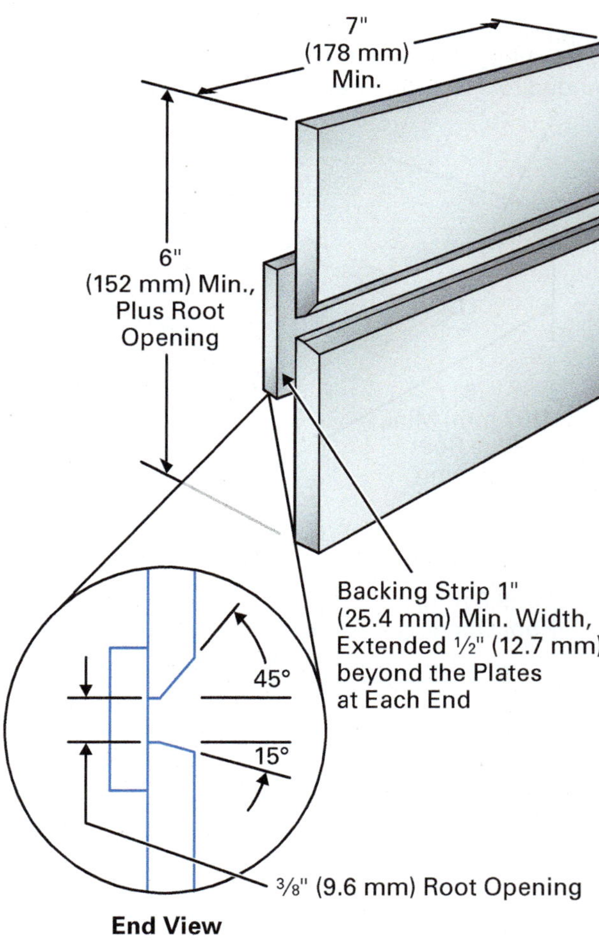

Figure 11 Alternate horizontal aluminum weld coupons.

1.0.0 Section Review

1. What heavy welding gas can concentrate in low areas and ignite or explode if exposed to sparks?
   a. Nitrogen
   b. Propane
   c. Compressed air
   d. Helium

2. The *proper* practice for aluminum is to use a contact tube on the GMAW gun that is _____.
   a. one size larger than the wire diameter
   b. one size smaller than the wire diameter
   c. designed for stainless steel but still fits the aluminum wire
   d. designed specifically for the aluminum filler wire to ensure proper sizing

3. What is the *negative* factor related to the use of helium in the shielding gas for aluminum?
   a. It does not prevent oxide formation very well.
   b. It significantly increases the required welding current.
   c. It is more expensive than argon.
   d. Some aluminum alloys are damaged by it.

2.0.0 GMAW Concepts and Welds

Objective

Describe GMAW concepts and how to produce fillet and V-groove welds.

a. Identify the types of aluminum alloys.
b. Identify key challenges related to aluminum welding.
c. Describe weld beads, terminations, and restarts.
d. Describe how to make aluminum fillet welds in various positions.
e. Describe how to make aluminum plate V-groove welds in various positions.

Performance Tasks

1. Complete the following GMAW aluminum plate welding tasks:
   - Stringer beads
   - Weave beads
   - Weld terminations
   - Weld restarts
   - Overlapping beads
2. Complete GMAW fillet welds on aluminum plate in the 1F, 2F, 3F, and 4F positions.
3. Complete GMAW V-groove welds on aluminum plate with backing in the 1G, 2G, 3G, and 4G positions.

GMAW is an arc welding process that uses a continuous, consumable, solid wire electrode for the filler metal and a shielding gas to protect the weld zone (*Figure 12*). The GMAW process is commonly used to make welds on aluminum, as well as carbon steel, low-alloy steel, and stainless steel. Typical GMAW air-cooled and water-cooled guns are shown in *Figure 13*.

GMAW is a fast and effective method for producing high-quality welds. Because this type of welding can be done continuously using a wire feeder, discontinuities and restarts are reduced.

For the purposes of this training module, the dimensions used are representative of various codes and standards and may not be specific to any particular code. Always refer to the applicable code, standard, or site WPS for the task at hand.

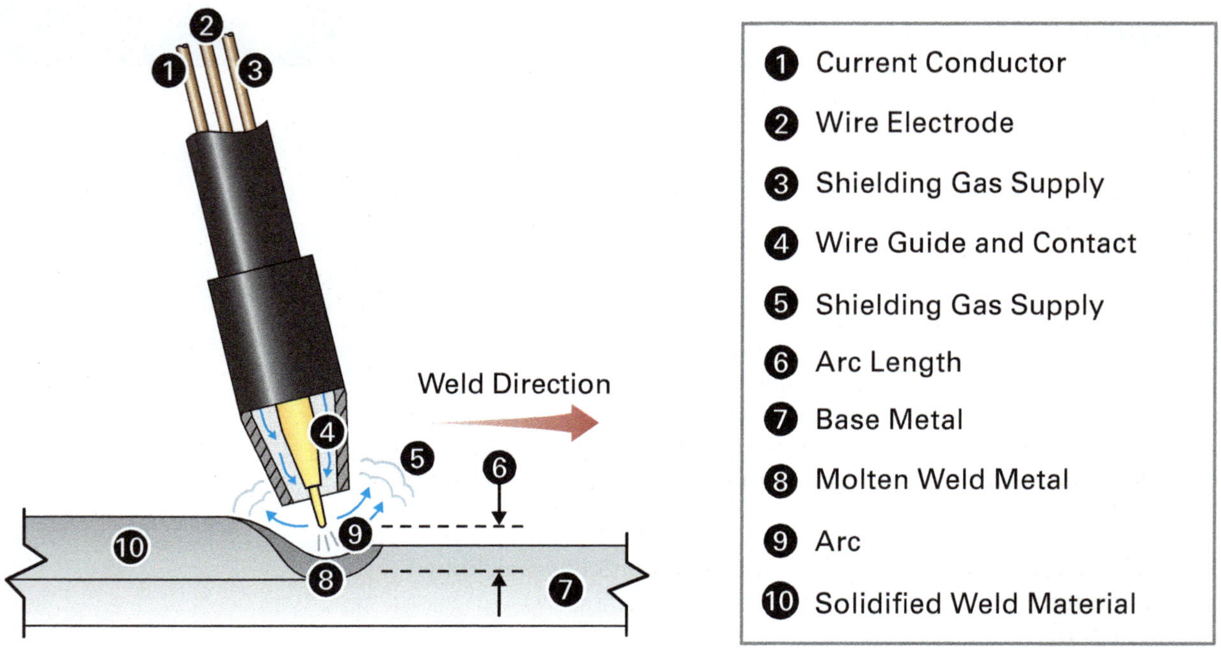

Figure 12 GMAW welding process.

(A) Inverter Welding Machine

(B) Air-Cooled Gun

(C) Water-Cooled Gun

Figure 13 Typical inverter-based GMAW equipment.
Source: Photo courtesy of The Lincoln Electric Company, Cleveland, OH, U.S.A.

2.1.0 Aluminum Alloys and Characteristics

Aluminum and aluminum alloys are often used in manufacturing because they are lightweight and have good corrosion resistance. Aluminum is one-third the weight of steel but is six times more thermally conductive. Pure aluminum is fairly soft, has low strength, and is not suitable for carrying significant loads.

Aluminum is nontoxic, nonsparking, and nonmagnetic. Molten aluminum has a silvery sheen under the glow of the welding arc. When heated, aluminum does not change color like steel does.

Aluminum is a highly reactive element. This reactivity explains why it is never found in nature in its free state, even though it is the most common metal on earth. Aluminum has a strong chemical attraction toward oxygen. This causes aluminum to oxidize and form aluminum oxide, the aluminum version of rust, as soon as it is exposed to air. Aluminum oxide is extremely hard and has a high melting point. Aluminum is also porous, allowing it to hold moisture and contaminants, which can result in weld porosity. This tendency is especially true of aluminum alloys containing magnesium.

Aluminum can be strengthened by adding small amounts of alloy metals to the base metal. Alloys are also used to enhance various other properties of aluminum. Alloys with common alloying elements are grouped together into a series. The alloys in each series are further defined by their ability to be treated by heat.

Heat-treatable and non-heat-treatable aluminum alloys each react differently to welding. Strain hardening is a method of increasing the strength of a metal by cold working it. Alloys that respond to cold working are said to be strain-hardenable, separating them from heat-treatable alloys. *Table 1* shows aluminum alloy characteristics by series, classification, and major alloying elements.

TABLE 1 Aluminum Alloy Characteristics

Series	Classification	Major Alloying Elements	Characteristics
1xxx 1050A	Non-heat-treatable	High purity with small amounts of iron and silicon	Relatively soft, low-strength, non-load-carrying; excellent corrosion resistance
2xxx 2014A	Heat-treatable	Copper	High strength; prone to liquation under welding heat
3xxx 3103	Non-heat-treatable	Manganese	Stronger and more ductile than 1xxx grades; excellent corrosion resistance
4xxx 4043A	Filler wires	Silicon	Excellent corrosion resistance
5xxx 5083	Non-heat-treatable	Magnesium	Good ductility; excellent corrosion resistance; good for situations involving impact and shock-loading; not recommended for service with prolonged exposure to high temperatures
6xxx 6082	Heat-treatable	Silicon and magnesium	Good as-welded strength; easily extruded; not recommended for exposure to electrolytes, such as seawater
7xxx 7020	Heat-treatable	Zinc	Good weldability and high as-welded strength
8xxx 8090	Non-heat-treatable and heat-treatable	Other	Not normally welded

Aluminum metallurgy can be divided into the following categories:

- Non-heat-treatable alloys
- Heat-treatable alloys
- Filler metal alloys
- **Cast aluminum**
- **Wrought aluminum**

Cast aluminum: Aluminum alloys that are molded into their final form by one of various methods such as sand-casting, die-casting, or pressure die-casting. Cast aluminum contains high levels of silicon to enhance the material's ability to form complicated castings.

Wrought aluminum: Aluminum alloys that are initially formed into ingots or similar shapes and are then worked into their desired shape through bending, rolling, extrusion, or other mechanical methods.

2.1.1 Non-Heat-Treatable Alloys

The strength of non-heat-treatable aluminum alloys depends on elements such as silicon (Si), iron (Fe), manganese (Mn), and magnesium (Mg). These alloying elements are found in the 1xxx, 3xxx, and 5xxx series of alloys. When welded, these alloys may lose the effects of work hardening, which results in softening of the heat-affected zone (HAZ) next to the weld.

Additional strengthening comes from strain hardening in the forming process, such as by cold rolling. Welds in non-heat-treatable alloys have excellent ductility and can withstand severe distortion before failing, and welds in heat-treatable alloys have lower ductility. The basic temper designations for non-heat-treatable and heat-treatable aluminum alloys, such as castings, are as follows:

- *O* — Annealed, recrystallized
- *F* — As fabricated
- *H1(x)* — Strain hardened only where $x = 2, 4, 6, 8$, or 9, representing ¼, ½, ¾, full hardness, or extra hardness
- *H2(x)* — Strain hardened and then partially annealed where $x = 2, 4, 6, 8$, or 9, representing ¼, ½, ¾, full hardness, or extra hardness
- *H3(x)* — Strain hardened and then stabilized where $x = 2, 4, 6, 8$, or 9, representing ¼, ½, ¾, full hardness, or extra hardness
- *W(HR)* — Solution heat treated where *HR* = fractional and full hours of aging at room temperature after heat treatment
- *T(N)* — Thermally treated to produce tempers other than *O*, *F*, or *H* where *N* = 1 through 10, which indicates a specific treatment, sometimes followed by another digit representing a variation of the treatment

Silicon

One of the advantages to having silicon in aluminum alloys is that silicon expands as it cools. This characteristic helps prevent cracking in aluminum-alloy welds as the silicon expands to help fill spaces that might otherwise be subject to solidification or liquation cracking.

2.1.2 Heat-Treatable Alloys

Heat-treatable alloys contain copper, magnesium, zinc, or silicon. These elements become more soluble when heated. Heat-treatable aluminum alloys develop their properties through **solution heat treatment** and quenching, followed by either **natural aging** or **artificial aging**.

Solution heat treatment: A process applied to a heat-treatable aluminum alloy that raises the temperature of the alloy to roughly 1,000°F (538°C) for a set period of time, causing the alloy elements to disperse more consistently throughout the metal. The metal is then quenched with water to ensure the dispersed elements remain in place within the metal structure rather than precipitate out.

Natural aging: A process that follows solution heat treatment and other processes that allows the alloy to simply rest at room temperature for a prescribed period of time.

Artificial aging: A process following solution heat treatment and other processes that allows the alloy to rest for a prescribed period of time at an elevated temperature.

Welding heat applied over an extended period reduces the strength of heat-treatable aluminum alloys. As a result, factors such as welding speed and cooling rate have an influence on the final properties of an alloy. Because of the time factor, the HAZ in heat-treatable alloys is usually only partially annealed. Welding redistributes the hardening components in the HAZ, reducing strength locally (close to the weld). The 2xxx, 6xxx, 7xxx, and some of the 8xxx series contain the principal heat-treatable alloying elements. The heat-treatable temper designations include O, F, W, and T1 through T10.

Postweld heat treatment (PWHT) increases the strength of the HAZ, but it also reduces the amount of ductility. Strength improvement depends somewhat on the filler metal used. Preheating is seldom recommended for welding heat-treatable alloys.

Joint thickness can also have an effect on weld properties in heat-treatable alloys. Thinner metal can be welded with a minimum of heat input and with a high cooling rate. This results in higher as-welded strength. For thicker metal, a number of stringer beads are preferable to a few large beads.

2.1.3 Cast Aluminum

Cast aluminum has high silicon content for increased fluidity. It does not anodize as well as wrought aluminum. Cast aluminum is identified by three-digit grade numbers, such as 206, 319, and 356. *Table 2* lists the numbers representing the alloying elements used with cast aluminum. The first number indicates the alloying element. The second and third numbers represent the specific alloy by formula. The number after the decimal point shows whether the alloy composition is for the final casting (.0) or for the **ingot** (.1 or .2, depending on purity). Modifications to cast aluminum are shown by a letter in front of the number. An example of a typical silicon-magnesium alloy for final casting is A356.0. The same alloy in ingot form would be A356.1 or A356.2.

Ingot: A mass of metal shaped for convenient storage and transport.

Cast Aluminum

Cast aluminum alloys are given their final form by one of several methods. Sand-casting is a popular process where a mold is made of sand that can be reused numerous times. Molten aluminum is then poured into the mold. Pressure die-casting processes use molds made of steel, and the molten aluminum is injected into the mold under pressure. Casting is best for complex shapes, such as the one shown below.

Source: iStock

TABLE 2 Alloy Numbers for Cast Aluminum

Alloy Series	Principal Alloying Elements
1xx.x	Essentially pure aluminum
2xx.x	Copper
3xx.x	Silicon plus copper and/or magnesium
4xx.x	Silicon
5xx.x	Magnesium
6xx.x	Unused series
7xx.x	Zinc
8xx.x	Tin
9xx.x	Other elements

2.1.4 Wrought Aluminum

Wrought aluminum alloys are identified using four numbers. *Table 3* shows the alloy series numbers and the principal alloying elements used with the various wrought aluminum alloys. The first number indicates the principal alloying element or elements. The second number signifies some change to the alloy (a zero indicates that there was no change). The third and fourth numbers identify the specific alloy by formula. For example, Alloy 5054 is an aluminum-magnesium alloy without modifications. It has the assigned alloy number of 54. Alloy 5154 is the same alloy after modification. The only exception is the 1xxx series, which is for pure aluminum. In that series, the last two numbers show a degree of purity above 99%. For example, Alloy 1060 has a purity of 99.6%.

Most of the wrought grades in the 1xxx, 3xxx, 5xxx, 6xxx, and the medium-strength 7xxx series (for example, 7020) can be fusion welded using both the GTAW and GMAW processes. The 5xxx series alloys in particular have excellent weldability. Wrought aluminum is commonly formed into plates or sheets, rods, bars, wire, and extruded products such as tubing.

TABLE 3 Alloy Numbers for Wrought Aluminum

Alloy Series	Principal Alloying Elements
1xxx	Aluminum (99% minimum and greater)
2xxx	Copper
3xxx	Manganese
4xxx	Silicon
5xxx	Magnesium
6xxx	Magnesium and Silicon
7xxx	Zinc
8xxx	Other elements
9xxx	Unused series

Wrought Aluminum

Wrought alloys start out as cast ingots or billets (bars, blocks, or bricks of metal), which are then formed into the desired shape mechanically by rolling, extruding, or forging. The material may be formed hot or cold, depending on the requirements. The fencing material shown here is constructed of wrought aluminum.

Source: iStock

2.2.0 Aluminum Welding Challenges

There are several challenges associated with welding aluminum. This section focuses on some of the primary challenges related to aluminum welding. Preheating of aluminum is not usually required for GMAW. However, the high thermal conductivity of aluminum demands a high rate of heat input for welding. In some cases, very thick sections of aluminum may require preheating, typically at less than 250°F (121°C). Using a higher welding current, adding helium to the shielding gas mix, or more precisely controlling other welding variables are all ways to generate an appropriate amount of GMAW welding heat for aluminum.

Aluminum GMAW is done with direct current electrode positive (DCEP) constant-voltage units. When welding thick aluminum that requires deep penetration, a shielding gas of helium mixed with argon is sometimes used, but at a greater cost. Pulsed GMAW is used to reduce heat input and prevent burn-through in thin sections of aluminum.

Aluminum welding current is mainly related to thickness. This is because the current must be high enough to give the required penetration. If the weld puddle becomes too large, the welding speed may need to be increased beyond that comfortable for the welder. To prevent melt-through on thin materials and on root passes, backing strips or rings made of stainless steel or a nonmetallic material are sometimes used. The backing strips or rings are often made with a shallow groove in the surface. The groove ensures that penetrating welds have the required size and shape when cool. On advanced welding machines used for GMAW, pulsing control can also be used to help eliminate melt-through problems.

2.2.1 Surface Preparation

Aluminum picks up hydrogen gas from condensation, which can form either during a rapid rise in temperature or when the metal is cooled below the dew point of the environment. Because materials are often stored in unheated areas, special care should be taken with aluminum to raise its temperature slowly. Another recommendation is to store aluminum in the vertical position to allow any condensation to run off. To reduce the risk of condensation forming on the surface, allow aluminum to stabilize at room temperature for one hour before welding. When bringing cold material into a warmer, more humid environment, avoid welding it immediately.

Aluminum combines with oxygen from the atmosphere to form a surface film of very hard oxide. This process can even occur at room temperature. Pure aluminum melts at 1,220°F (660°C), but its oxide melts at 3,600°F to 3,900°F (1,980°C to 2,150°C). Aluminum oxide is harder than most of the substances it encounters and can damage a milling cutter if the oxides are trapped in the welds. Because of its hardness, aluminum oxide is often used for the grit in grinding wheels. Aluminum oxide also tends to absorb moisture (hygroscopic). Under the extreme heat of welding, the moisture breaks down to free hydrogen, which often causes porosity in the weld. As a safeguard against porosity, all oxidation must be thoroughly removed from the base metal before welding.

WARNING!

Grinding wheels used on aluminum must be made for aluminum only. They should not be used on any other type of metal or on surfaces contaminated with grease or oil. Using other types of grinding wheels on aluminum may cause the grinding wheels to shatter.

Proper cleaning of welding surfaces cannot be stressed enough. Oxides, greases, and oil films can act as contaminants and cause unsound welds. Aluminum should be welded as soon as possible after cleaning or within one half hour of cleaning. If a delay occurs, the weld zone may have to be cleaned again. If oxides are not removed, the weld will be subject to poor fusion and inclusions. This poor outcome is due to the reduced wetting action of the molten weld pool as the aluminum comes in contact with any remaining oxide-contaminated weld joint surfaces.

Aluminum should be hand brushed between passes with a stainless steel brush exclusively used on aluminum. Make sure that the brush has not been used on other materials or for any other purpose. Water, oil, and grease represent the contaminants generally found on aluminum. Wire brushing before these contaminants are removed only pushes them into the surface or damages the aluminum. Oil and grease must be removed before cleaning aluminum with a wire brush.

Both water-based aluminum deoxidizers and solvent cleaners such as acetone can be used to remove oil and grease. When using water-based cleaners, all water must be removed before welding. Most hydrocarbon-based solvents are highly volatile and evaporate quickly. Water-based cleaners, however, must be thoroughly wiped away or heat dried before welding. Review the Safety Data Sheet (SDS) for any additional safety hazards or precautions associated with these cleaners.

Prewelding cleaning solutions, also referred to as *chemical cleaners*, can remove both the aluminum oxide and any contaminants, but it does carry risks. The disadvantage of prewelding is that deoxidizers for aluminum are very aggressive chemicals. These chemicals usually contain dilute acids that can etch the surface. They can be dangerous to the people using them as a result. It is very important to follow the manufacturer's instructions and take the necessary safety precautions. Always review and follow the recommendations of the SDS when using cleaning solvents. When using deoxidizers, the metal must be completely rinsed and dried. This avoids creating a surface film of hydrated oxide.

WARNING!

Be very careful when using preweld chemical cleaning solutions. They can harm not only the environment, but also the people using them. Make sure that all cleaning solutions are properly contained. Always be sure to wear the appropriate PPE. Refer to the applicable SDS and the product label for instructions on how to properly dispose of any used cleaning solutions and the materials used to apply them.

2.2.2 Filler Metal Alloys

Filler metal composition is determined in part by the weldability and properties of the base metal. Other factors to consider include corrosion resistance, exposure of the weldment to sustained high temperatures, and the requirements for **anodic coating**. For non-heat-treatable alloys, matching fillers are often used. For low-alloy materials and heat-treatable alloys, nonmatching fillers are used to prevent solidification cracking.

Filler metal alloys ER4043 and ER5356 are used in roughly 85% of aluminum welds. They are also the most readily available filler metal products for aluminum. Your instructor will likely determine the use of a specific filler metal that is most appropriate for your practice coupons. When there is a WPS or site quality standard, the filler rod type will be specified. It will be given as an AWS classification and/or manufacturer's standard. It will sometimes include a manufacturer's name. When a WPS for the weld exists, the filler metal that is specified must be used.

Anodic coating: An artificial buildup of aluminum oxides on the surface of aluminum alloys that improves corrosion resistance. This coating is many times thicker than a coating of naturally occurring oxides.

CAUTION

Filler metal specifications in the WPS or site quality standards must be followed. Failure to use the specified filler metal will result in a weld quality issue and the potential for rejection.

Aluminum filler rods for GTAW are usually selected for their compatibility with the chemical composition of the base metal. There are many distinct types of aluminum base metals and filler metals. *AWS Standard A5.10/A5.10M:2017, Welding Consumables—Wire Electrodes, Wires and Rods for Welding of Aluminum and Aluminum-Alloys—Classification,* provides a guide for the selection of bare aluminum filler metals for specific aluminum base metals and alloys. Additionally, charts and tables provided by filler metal manufacturers assist in selecting the correct filler metal from their product line. An example of a filler metal selection chart from Hobart Brothers Company is provided on the NCCER Digital Resources page, which can be accessed using the QR code at the beginning of this module.

High-quality aluminum welds can only be achieved if the filler metal used is clean and of high quality. Contaminated filler metal transfers the contamination to the weld pool. Filler metal must be used as soon as possible after removal from the shipping or storage packaging. When this is not possible, the filler metal must be stored in airtight storage containers to prevent further oxidation. Also remember that heating aluminum and other metals accelerates the oxidation process dramatically. This is especially true when the humidity is high.

Manufacturers' Data for Filler Metals

Most manufacturers of filler metal publish paperback welding guides that include information about their filler materials. These booklets can be obtained through your distributor for free or for a small charge. An increasing number of manufacturers also offer apps for filler metal selection that can be used on smartphones, tablets, and similar devices.

2.2.3 Porosity, Cracking, and Weld Profile Defects

Before fusion welding can take place, all aluminum oxide must be removed. Remember that the melting point of aluminum oxide is about three times that of aluminum. Because of this temperature disparity, the welding heat needed to break down the oxide causes the aluminum to melt long before the oxide does and **coalescence** cannot take place. Complete removal of the oxide prior to welding is essential to avoid this problem.

Coalescence: The growing together or growth into one body of the materials being joined.

The following are the most common problems in aluminum welding:

- Porosity
- Cracking
- Poor weld bead profile

Table 4 shows the causes, contributing factors, and corrective measures related to porosity and cracking in GMAW. These issues are explored in greater depth in the sections that follow.

Porosity

Aluminum welds are highly prone to porosity. The main cause of porosity is the absorption of hydrogen into the weld pool. The hydrogen forms separate pores in the weld metal as the weld pool solidifies. Some common sources of hydrogen are hydrocarbons and the atmosphere. Another is the moisture from contaminants on the parent material and on the surface of the filler wire. In some cases, a torch gas flow that is set too high can cause porosity in the weld. It does this by creating turbulence that draws air into the gas, causing porosity. *Figure 14* shows porosity caused by contamination in a cast aluminum alloy 356 weld.

Thoroughly clean the surface to reduce the risk of porosity. Proper cleaning techniques include solvent degreasing and chemical etching. Surface cleansing should be followed by mechanical cleaning by grinder buffing or wire brushing.

TABLE 4 Causes and Cures for Porosity and Cracking in GMAW

Cause	Contributing Factors	Corrective Measures
Hydrogen	Hydrogen is the result of dirt-containing oils or other hydrocarbons; moisture in atmosphere or on metal; a hydrated oxide film on metal; or moisture in gas or gas lines. Base metal may be the source of entrapped hydrogen (the thicker the metal, the greater the possibility of hydrogen). Spatter contributes to hydrogen formation.	Degrease and mechanically or chemically remove oxide from weld area. Avoid humidity; use dry metal or wipe dry. Reduce moisture content of gas. Check gas and water lines for leaks. Increase gas flow to compensate for increased hydrogen in thicker sections. To minimize spatter, adjust welding conditions.
Impurities	Cleaning or other compounds, especially those containing calcium.	Use recommended cleaning compounds; keep work free of contaminants.
Incomplete root penetration	Insufficient welding current or preheat.	Use a higher welding current or redesign the joint geometry. Preheat.
Temperature	Running too cool tends to increase porosity that is caused by premature solidification of molten metal.	Maintain the proper current, arc length, and torch travel speed relationship.
Welding speed	Too great a welding speed may increase porosity.	Decrease the welding speed and establish and maintain proper arc length and current relationship.
Solidification time	Quick freezing of the weld pool traps any gases present, causing porosity.	Establish the correct welding current and speed. If the work is appreciably below room temperature, use supplemental heating.
Chemical composition of weld materials	Pure aluminum weld metal is more susceptible to porosity than an aluminum alloy.	If the porosity is excessive, try an alloy filler material.
Cracking	Such causes of porosity as temperature, welding time, and solidification may also be contributing causes of cracking. Other causes may be discontinuous welds, welds that intersect, repair welds, cold working either before or after welding, and weld-metal composition. In general, crack-sensitive alloys include those containing 0.4% to 0.6% silicon,1.5% to 3.0% magnesium, or 1.0% copper.	Lower current and faster speeds often prevent cracking. However, a change to a filler alloy that brings the weld-metal composition out of the cracking range is recommended where possible.

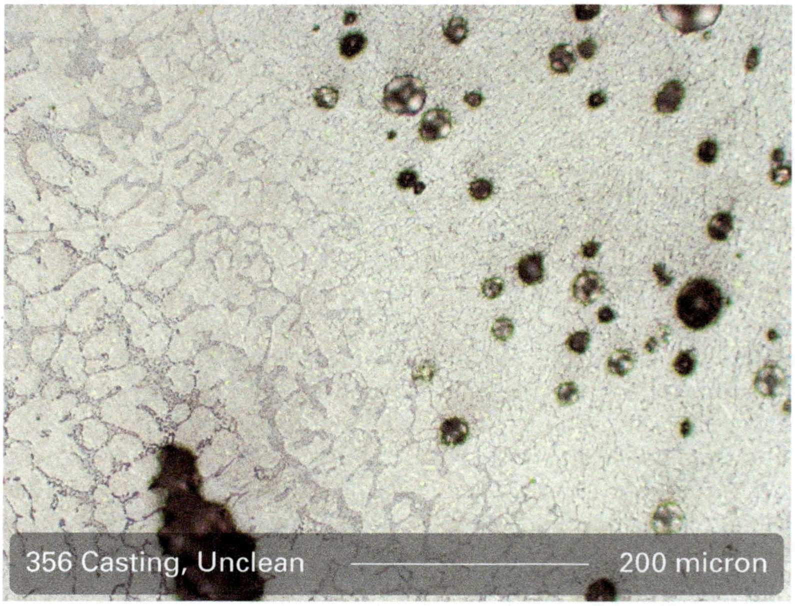

Figure 14 Aluminum porosity example.
Source: Microlabgallery.com

Cracking

Cracking occurs in aluminum alloys because of high stress. The stresses are generated across the weld by the high degree of thermal expansion, which is twice that of steel. Thermal expansion is the fractional increase in size due to a rise in temperature. Cracking can also be traced to the amount of contraction that occurs when the weld cools and solidifies. Contraction is usually 5% more in aluminum than in similar welds on steel. High-strength aluminum alloys (such as 7010 and 7050) and most of the 2xxx series of alloys are not recommended for fusion welding. This is because welding these alloys commonly results in both **solidification cracks** and **liquation cracks**. However, these types of cracks can also occur in other alloys.

Solidification cracks: Cracks that occur as the filler and base metal solidify, when the remaining liquid weld metal is unable to fill the spaces between areas of solidifying weld metal. Both solidification and liquation cracking may be referred to as *hot cracking*.

Liquation cracks: Liquation is the process of separation by melting, such as in an alloy when one element melts while the others remain solid. Liquation cracks result from localized melting in areas such as the grain boundaries, while adjacent grains remain solid. Both solidification and liquation cracking may be referred to as *hot cracking*.

Solidification cracks (*Figure 15*), also referred to as *hot cracks*, form in the center of the weld and run along its centerline. Solidification cracks also occur in the weld crater at the final termination. There are two main causes of solidification cracks. One is the incorrect combination of filler wire and base metal. The other is welding under high-restraint conditions.

Figure 15 Solidification crack in an aluminum weld.
Source: Image copyright TWI Ltd, reproduced by permission

The use of a nonmatching, crack-resistant filler metal (usually from the 4xxx and 5xxx series alloys) can reduce the risk of solidification cracking. However, the use of these fillers can also result in a weld of lower strength, and the weld may not respond to heat treatment. The weld bead must be thick enough to withstand the stresses caused by contraction as the weld solidifies.

There are several ways to reduce the degree of restraint on the weld. Methods include correct edge preparation, accurate joint setup, more compatible filler alloy, and the correct weld sequence. Higher weld speeds that reduce the size of the HAZ can also reduce solidification cracking.

Preheating reduces stress on the solidifying weld metal. It does this by reducing the temperature change across the HAZ and by permitting faster joining. However, to reduce the risk of stress cracking, preheating should only be used when the joint is unrestrained. Preheating should be limited to temperatures under 250°F (121°C) to preserve the mechanical properties of the base metal.

Liquation cracks occur in the HAZ at the grain boundaries when low-melting films are formed. These films are not strong enough to withstand the contraction stresses created when the weld metal solidifies and cools. Heat-treatable alloys, such as the 6xxx, 7xxx, and 8xxx series, are more likely than other alloys to crack in this way.

Use of a filler metal that has a lower melting temperature than the base metal can reduce the risk of liquation cracking. An example would be to use a 4xxx filler metal to weld a 6xxx-series-alloy base metal. However, 4xxx filler metal should not be used to weld high-magnesium alloys, such as 5083. This is because excessive magnesium silicide may form at the fusion boundary, causing a loss of ductility and an increased risk of cracking.

Poor Weld Bead Profile

A poor weld bead profile can result from setting incorrect welding parameters. It can also be traced to poor welding techniques. Some common weld profile defects are lack of fusion, lack of penetration, and undercut. Aluminum alloys are especially prone to profile imperfections. This tendency is due to their high thermal conductivity and the rapidly solidifying weld pool.

2.3.0 Beads, Terminations, and Restarts

The two basic bead types are stringer beads and weave beads. Aluminum welds are easily contaminated. Because of this, extreme care must be taken in cleaning the base metals just before starting either type of bead. Always refer to the WPS for cleaning instructions and for the proper chemicals to use to remove oil and grease.

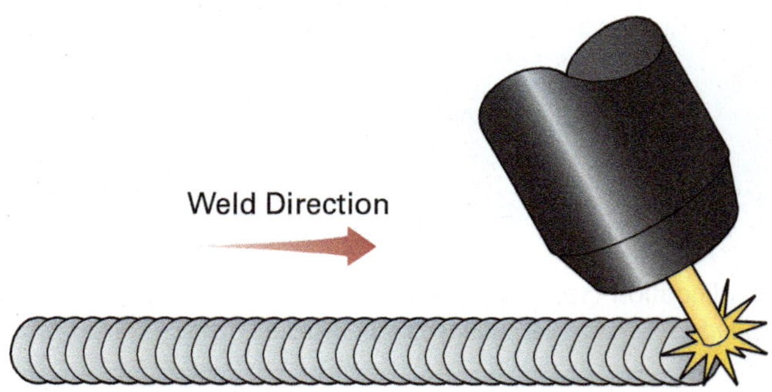

Figure 16 Stringer bead.

2.3.1 Stringer Beads

Stringer beads (*Figure 16*) are made with little or no side-to-side motion of the gun. Practice running stringer beads in the flat position. Experiment with different push angles and electrode stickouts.

Observe the following steps to make stringer beads:

Step 1 Clean the base metal thoroughly before welding.

WARNING!

To prevent personal injury when using cleaning solvents, always comply with the SDS guidelines for the cleaning solvents being used.

Step 2 Hold the gun at the desired angle with the electrode tip directly over the point where the weld will begin. Pull the gun trigger.

Step 3 Hold the arc in place until the weld puddle begins to form.

Step 4 Slowly advance the arc while maintaining the proper work and travel angles. Stay in the leading edge of the puddle to maintain cleaning action and to ensure fusion.

Step 5 Stop forward movement and back up about ⅛" (3.2 mm).

Step 6 Release the trigger once the crater has been filled.

Step 7 Inspect the bead to ensure it exhibits the following characteristics:

- Straightness
- Uniform appearance
- Smooth, flat transition with complete fusion at the toes of the weld
- Crater filled
- No excessive porosity with no pores larger than $\frac{3}{32}$" (2.4 mm)
- No undercut greater than $\frac{1}{32}$" (0.8 mm) deep or 10% of the base metal thickness, whichever is less
- No underfill
- No cracks

Step 8 Continue practicing stringer beads until you can make acceptable welds every time.

2.3.2 Weave Beads

The side-to-side motions of the electrode used to make weave beads are called *oscillations* (*Figure 17*). The width of a weave bead is determined by the amount of side-to-side motion.

When making a weave bead, take special care at the toes to ensure proper fusion to the base metal. Do this by slowing down or pausing slightly at the edges. Pausing at the edges will also reduce undercut and give the weld the proper profile.

Practice running weave beads in the flat position. Experiment with different weave motions, push angles, and stickouts.

Use the following steps to make weave beads:

Step 1 Clean the base metal before welding.

Step 2 Hold the gun at the desired angle with the electrode tip directly over the point where the weld will begin. Pull the gun trigger.

Step 3 Hold the arc in place until the weld puddle begins to form.

Step 4 Slowly advance the arc in a weaving motion while maintaining the proper work and travel angles.

Step 5 Finish the bead by pausing until the crater has been filled, and then release the trigger.

Step 6 Inspect the bead to ensure it exhibits the following characteristics:

- Straightness
- Uniform appearance
- Smooth, flat transition with complete fusion at the toes of the weld
- Crater filled
- No excessive porosity with no pores larger than $\frac{3}{32}$" (2.4 mm)
- No overlap
- No underfill
- No undercut greater than $\frac{1}{32}$" (0.8 mm) deep or 10% of the base metal thickness, whichever is less
- No cracks

Step 7 Continue practicing weave beads until you can make acceptable welds every time. Clean the beads between each pass.

CAUTION

Do not exceed the recommended weave bead width, which is often specified in the welding code or WPS used at your site.

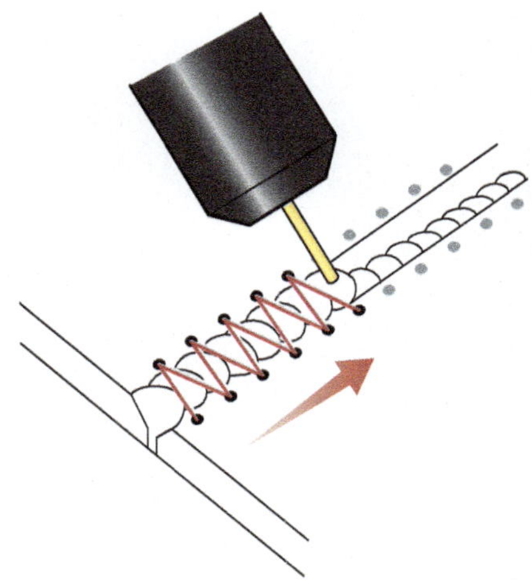

Figure 17 Weave bead motions.

2.3.3 Overlapping Beads

Overlapping beads are made by depositing connective weld beads parallel to one another. Each successive bead overlaps the previous bead to form a raised, flat surface. This is also called *padding*. Overlapping beads are used to build up a surface and to make multipass welds. Both stringer and weave beads can overlap. Properly overlapped beads, when viewed from their ends, form a nearly flat, level surface (*Figure 18*).

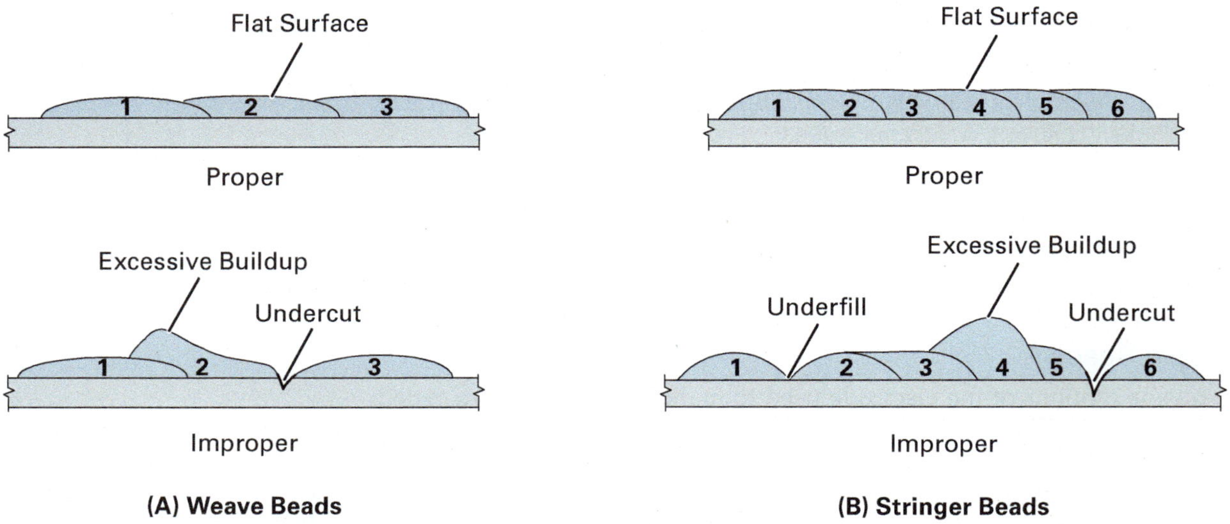

Figure 18 Proper and improper overlapping beads.

Follow these steps to weld overlapping stringer or weave beads using aluminum wire electrodes and the appropriate shielding gas:

Step 1 Mark out a square on a piece of aluminum.

Step 2 Clean and deoxidize the base metal.

Step 3 Weld a stringer or weave bead along one edge.

Step 4 Clean the bead with a stainless steel brush.

Step 5 Position the gun at a work angle of 10° to 15° toward the side of the previous bead to get proper fusion. Pull the trigger.

Step 6 Keep running overlapping stringer or weave beads until the square has been covered.

Step 7 Continue building layers of stringer or weave beads, one on top of the other, until your technique has been perfected.

2.3.4 Weld Terminations

A weld termination normally leaves a crater. Most welding codes require the crater be filled to the full cross section of the weld when making a termination. This can be difficult because most terminations are at the edge of a plate where welding heat tends to build up. This makes it harder to fill the crater.

To terminate a weld, use the following steps. The first three steps are further illustrated in *Figure 19*:

Step 1 Start to bring the gun up to a 0° travel angle.

Step 2 Slow the forward travel as you approach the end of the weld.

Step 3 Stop the forward movement and back up about ⅛" (3.2 mm).

Step 4 Release the trigger when the crater has been filled.

Step 5 Inspect the termination. The crater should be filled to the full cross section of the weld.

CAUTION

Do not immediately remove the gun from the weld until the puddle has solidified. After welding has stopped, the shielding gas post-flow that continues protects the molten aluminum. If the gun and shielding gas are withdrawn before the crater has solidified, contamination and defects can occur in the crater.

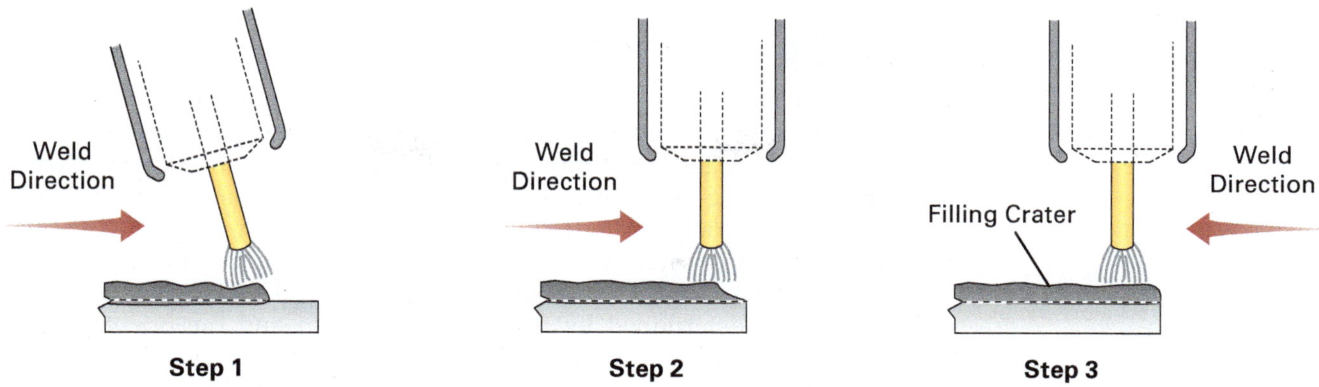

Figure 19 Terminating a weld.

Alternate Crater-Filling or Elimination Techniques

One way to eliminate the crater when terminating a GMAW weld on aluminum is to use run-off tabs. In cases where tabs cannot be used or are impractical, such as with pipe welds, you can reduce the weld pool before extinguishing the arc by increasing the travel speed just before releasing the gun trigger. This results in a small crater that may be free of cracks. Making several quick arc starts into the crater as it cools is another technique for filling the crater without adding heat to increase its size. Discuss these techniques with your instructor.

2.3.5 Weld Restarts

A restart is the junction where a new weld connects to and continues the bead of a previous weld. A restart must blend smoothly with the previous weld and must not stand out or create a weld defect. The technique for making a GMAW restart is the same for both stringer and weave beads.

Follow these steps to make a restart:

Step 1 Clean the base metal before welding.

Step 2 Hold the gun at the proper angle while restarting the arc directly over the center of the crater. Remember that arc strikes outside of the weld area are not acceptable.

Step 3 Move the electrode tip in a small circular motion over the crater to fill it with a molten puddle.

Step 4 Move to the leading edge of the puddle. As soon as the puddle fills the crater, continue the stringer or weave bead pattern.

Step 5 Inspect the restart.

A properly made restart blends into the bead and is hard to detect. If the restart has undercut, not enough time was spent in the crater to fill it. If the undercut is on one side or the other, use more of a side-to-side motion as you move back into the crater. If the restart has a lump, it was overfilled. It is likely that too much time was spent in the crater before the forward motion was resumed. On restarts, start the weld just beyond the crater, establish the arc, and move back toward the crater. This will prevent the development of a lump in this area.

2.4.0 Fillet Welds

Fillet welds require little base metal preparation, since there are no bevels, root faces, or root openings involved. However, the weld area must still be thoroughly cleaned, and any rough edges must be removed. Any dross from thermal cutting will cause porosity in the weld. For this reason, codes require that this material be removed entirely prior to welding. As noted previously, thermal cutting is not recommended for aluminum coupons.

The most common fillet welds are made in lap and T-joints. The weld position for plate is determined by the axis of the weld and the orientation of the workpiece. The positions for fillet welding on plate (*Figure 20*) are flat (1F), horizontal (2F), vertical (3F), and overhead (4F). In the 1F and 2F positions, the weld axis can be inclined up to 15°. Any weld axis inclination for the other positions varies with the rotational position of the weld face as specified in the AWS standards.

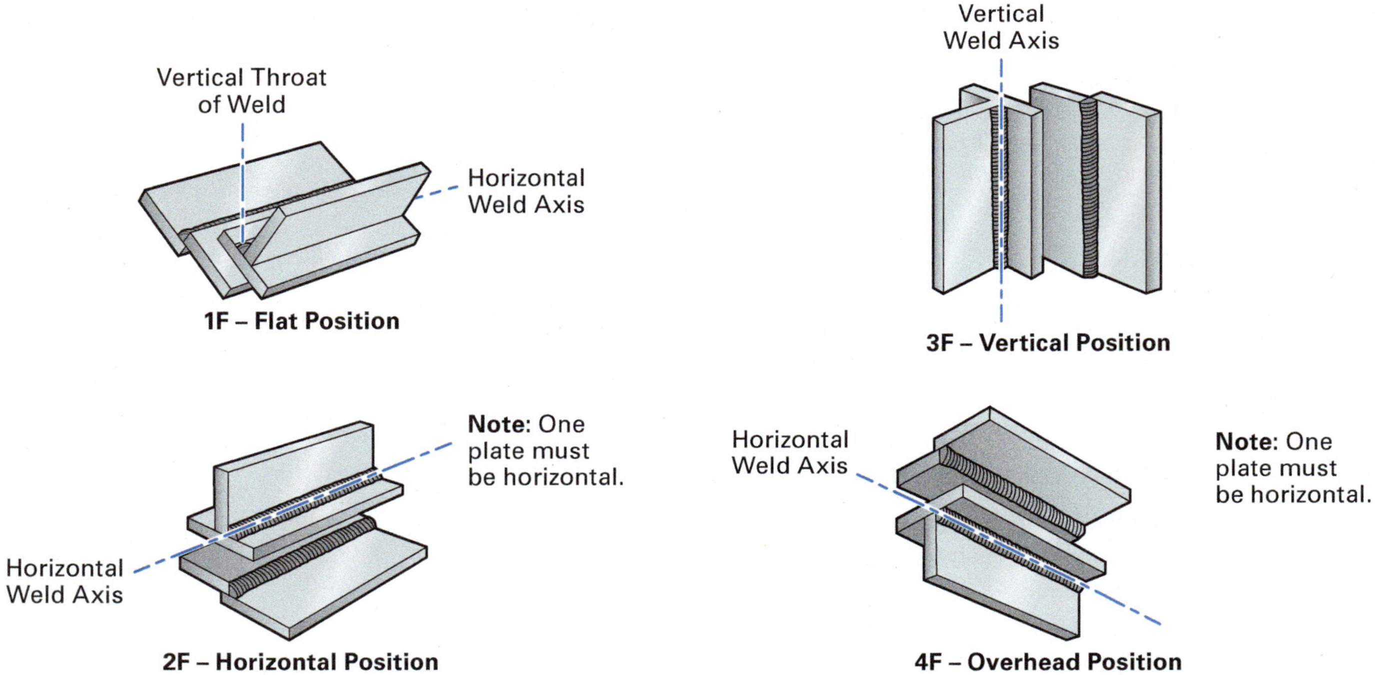

Figure 20 Fillet welding positions.

Fillet welds can be concave or convex, depending on the WPS or site quality standards. Welding codes require a fillet weld to have a uniform concave or convex face, although a slightly nonuniform face is acceptable. The convexity of a fillet weld or individual surface bead must not exceed that permitted by the applicable code or standard. A fillet weld with profile defects is unacceptable and must be repaired (*Figure 21*).

WARNING!

Unlike steels, aluminum does not change color when heated, and due to its high thermal conductivity, aluminum will get very hot quickly and not be visibly noticeable. To prevent injury, wear gloves when handling practice coupons and be sure to use pliers when the coupons are hot. Additionally, *OSHA Standard 29 CFR 1910.252(b)(4)(vii)* requires that welders, at the conclusion of welding operations, provide a warning sign for hot metal by marking the hot metal or providing some other means of warning other workers.

2.4.1 Practicing Flat (1F) Position Fillet Welds

Practice flat (1F) position fillet welds by making multipass (six-pass) convex fillet welds in a T-joint. Use an appropriate aluminum filler metal as directed by your instructor. When making flat fillet welds, pay close attention to the work angle and travel speed. For the first bead, the gun work angle is vertical (45° to both plate surfaces). Adjust the work angle for all subsequent beads (*Figure 22*). Clean each completed bead with a stainless steel wire brush before starting the next bead.

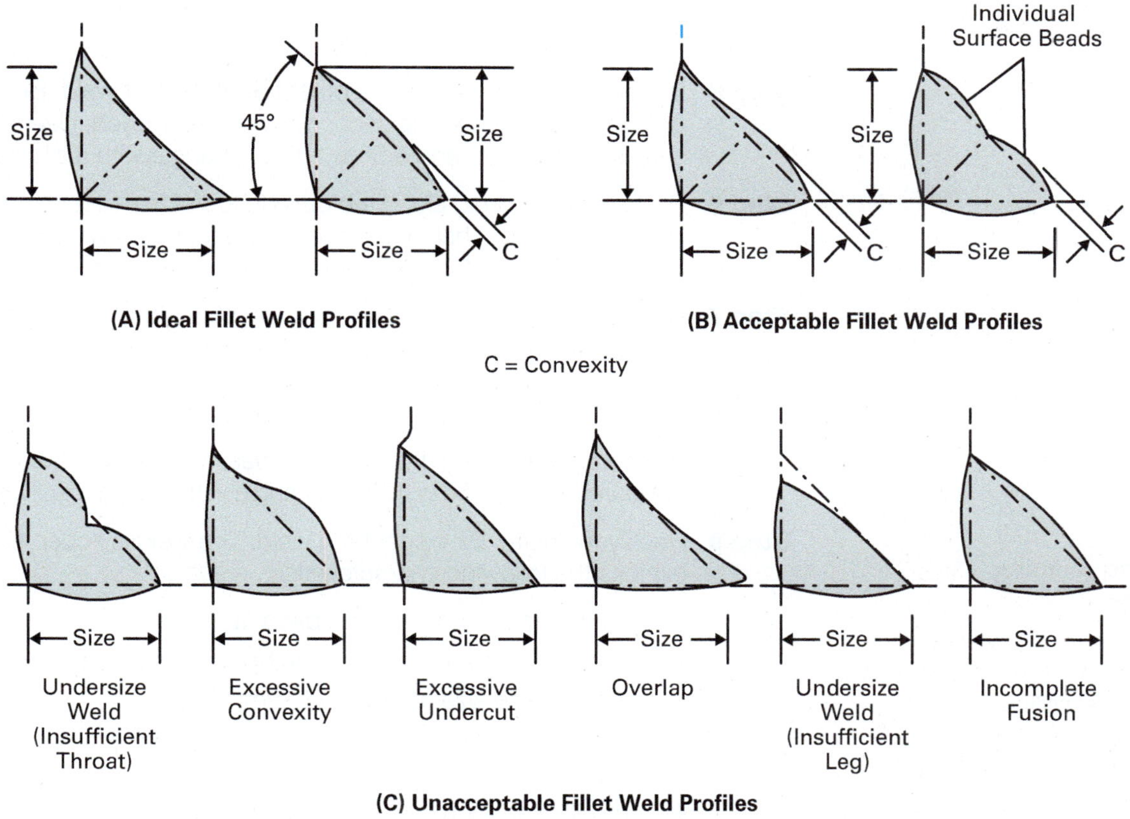

Figure 21 Acceptable and unacceptable fillet weld profiles.

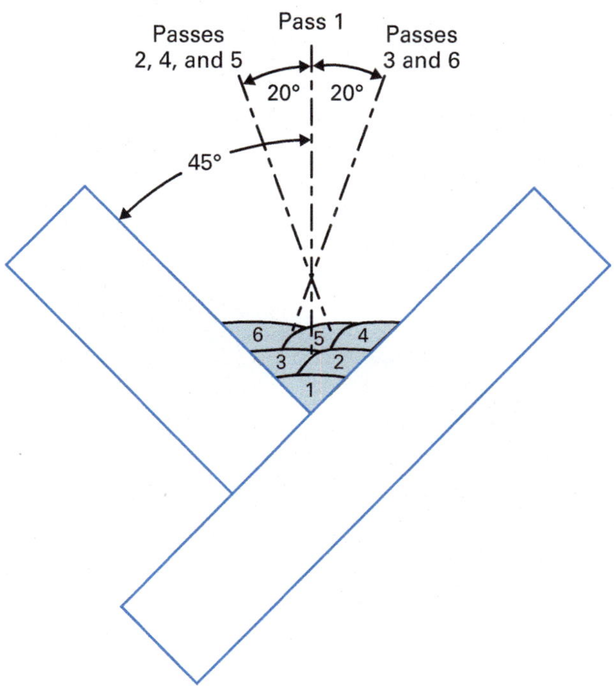

Figure 22 Multipass 1F bead sequence and work angles.

NOTE

In the following steps, clean all weld beads before beginning the next bead.

Follow these steps to make a flat fillet weld:

Step 1 Tack two plates together to form a T-joint for a fillet weld coupon (*Figure 23*). Clean the tack welds.

Step 2 Clamp or tack-weld the coupon in the flat (1F) position.

Step 3 Run the first bead along the root of the joint using a 45° work angle and a 5° to 10° push angle.

Step 4 Run the second bead along a toe of the first weld, overlapping about 75% of the first bead. Alter the work angle as shown in *Figure 22*. Use a 5° to 10° push angle with a slight oscillation.

Step 5 Run the third bead along the other toe of the first weld. Be sure to fill the groove created when the second bead was applied. Use the work angle shown in *Figure 22* and a 5° to 10° push angle with a slight oscillation.

Step 6 Run the fourth bead along the outside toe of the second weld, overlapping about half of the second bead. Use the work angle shown in *Figure 22* and a 5° to 10° push angle with a slight oscillation.

Step 7 Run the fifth bead along the inside toe of the fourth weld, overlapping about half of the fourth bead. Use the work angle shown in *Figure 22* and a 5° to 10° push angle with a slight oscillation.

Step 8 Run the sixth bead along the toe of the fifth weld. Be sure to fill the groove created when the fifth bead was run. Use the work angle shown in *Figure 22* and a 5° to 10° push angle with a slight oscillation.

Step 9 Have your instructor inspect the weld. The weld is acceptable if it exhibits the following characteristics:

- Uniform appearance on the bead face
- Craters and restarts filled to the full cross section of the weld
- Uniform weld size of ±$^1/_{16}$" (1.6 mm)
- Acceptable weld profile in accordance with the applicable code or standard
- Smooth transition with complete fusion at the toes of the weld
- No excessive porosity with no pores larger than $^3/_{32}$" (2.4 mm)
- No undercut greater than $^1/_{32}$" (0.8 mm) deep or 10% of the base metal thickness, whichever is less
- No overlap
- No underfill
- No inclusions
- No cracks

Tacking and Aligning Workpieces

When tacking workpieces together, position them to minimize distortion when the final welds are made. To achieve this, both sides of the workpieces are usually tacked with welds that are about $^1/_2$" (12.7 mm) long. After the first tack weld, use a hammer or other tool to align the workpieces side to side and end to end. Then tack the opposite side. Tack the far ends of the workpieces in the same way. Intermediate tack welds can be made every 5" to 6" (13 cm to 15 cm) as needed to minimize longitudinal distortion.

Figure 23 Fillet weld coupon.
Source: Terry Lowe

2.4.2 Practicing Horizontal (2F) Position Fillet Welds

Practice horizontal (2F) fillet welding by placing multipass fillet welds in a T-joint. Use an aluminum filler metal as directed by your instructor. When making horizontal fillet welds, pay close attention to the work angle. For the first bead, the electrode work angle is 45°. The work angle is adjusted for all other welds.

Follow these steps to make a horizontal fillet weld:

Step 1 Tack two plates together to form a T-joint for the fillet weld coupon. Clean the tack welds.

Step 2 Clamp or tack-weld the coupon in the horizontal position.

Step 3 Run the first bead along the root of the joint using a work angle of about 45° and a 5° to 10° push angle (*Figure 24*).

Step 4 Clean the weld.

Step 5 Run the remaining passes at the proper work angles using a 5° to 10° push angle and a slight oscillation, as shown in *Figure 24*. Overlap each previous pass. Clean the weld after each pass.

Step 6 Have your instructor inspect the weld. The weld is acceptable if it has the following features:

- Uniform appearance on the bead face
- Craters and restarts filled to the full cross section of the weld
- Uniform weld size of ±1⁄16" (1.6 mm)
- Acceptable weld profile in accordance with the applicable code or standard
- Smooth transition with complete fusion at the toes of the weld
- No excessive porosity with no pores larger than 3⁄32" (2.4 mm)
- No undercut greater than 1⁄32" (0.8 mm) deep or 10% of the base metal thickness, whichever is less
- No overlap
- No underfill
- No inclusions
- No cracks

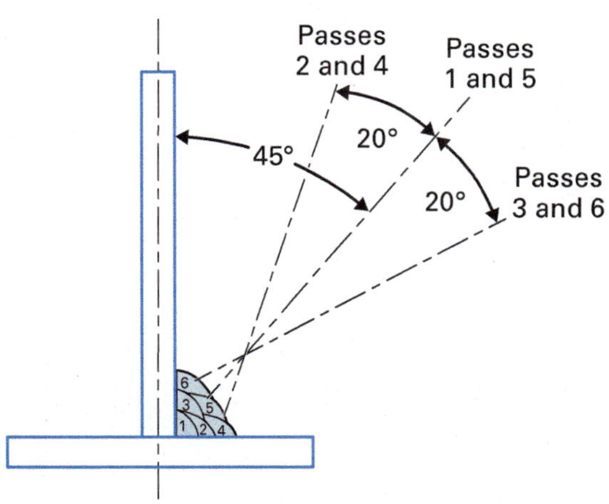

Figure 24 Multipass 2F bead sequence and work angles.

2.4.3 Practicing Vertical (3F) Position Fillet Welds

Practice vertical (3F) fillet welding by placing multipass fillet welds in a T-joint. Use an aluminum filler metal as directed by your instructor. Vertical welds are usually made by welding uphill from the bottom to the top using a push angle of 10° to 15°. Note that this push angle is slightly greater than the usual 5° to 10°. Because of the uphill welding and push angle, this type of welding is sometimes called vertical-up fillet welding. Either stringer or weave beads can be used for vertical welding. The site WPS or site quality standard will specify which technique to use on the job. Check with your instructor to see if you should practice stringer beads, weave beads, or both.

Follow the steps shown in *Figure 25* to make an uphill fillet weld using weave beads.

Step 1 Tack two plates together to form a T-joint for the fillet weld coupon.

Step 2 Clamp or tack-weld the coupon in the vertical position.

Step 3 Starting at the bottom, use a slight oscillating motion to run the first bead along the root of the joint. Use a work angle of about 45° and a 10° to 15° push angle. Pause in the weld puddle to fill the crater.

Step 4 Clean the weld.

Step 5 Run the remaining passes using a 10° to 15° push angle and a side-to-side weave technique with a 45° work angle. Use a slow motion across the face of the weld. Pause at each toe to penetrate and fill the crater. Clean the weld after each pass.

Step 6 Have your instructor inspect the weld. The weld is acceptable if it has the following features:

- Uniform appearance on the bead face
- Craters and restarts filled to the full cross section of the weld
- Uniform weld size of ±1⁄16" (1.6 mm)
- Acceptable weld profile in accordance with the applicable code or standard
- Smooth transition with complete fusion at the toes of the weld
- No excessive porosity with no pores larger than 3⁄32" (2.4 mm)
- No undercut greater than 1⁄32" (0.8 mm) deep or 10% of the base metal thickness, whichever is less
- No overlap
- No underfill
- No inclusions
- No cracks

Vertical Weave Bead

To help control undercut on a root pass, an alternate pattern, such as a small triangular weave, should be made. By pausing at the edges, the previous undercut greater than 1⁄32" (0.8 mm) deep or 10% of the base metal thickness, whichever is less can be filled. This action also creates undercut at the existing weld pool, but that will be filled in by the next weave. This technique should only be applied during the root pass.

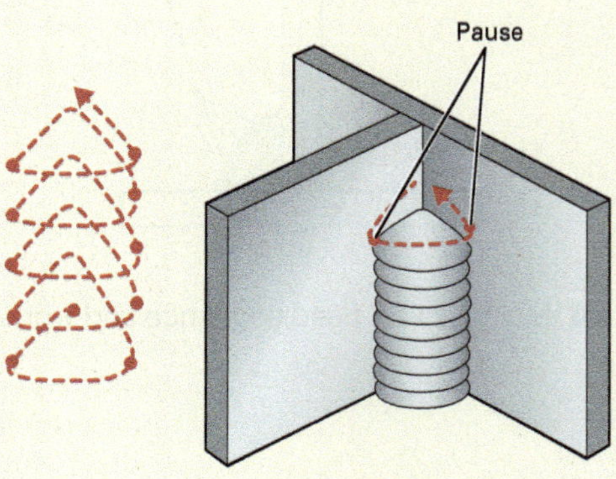

Triangular Weave Pattern

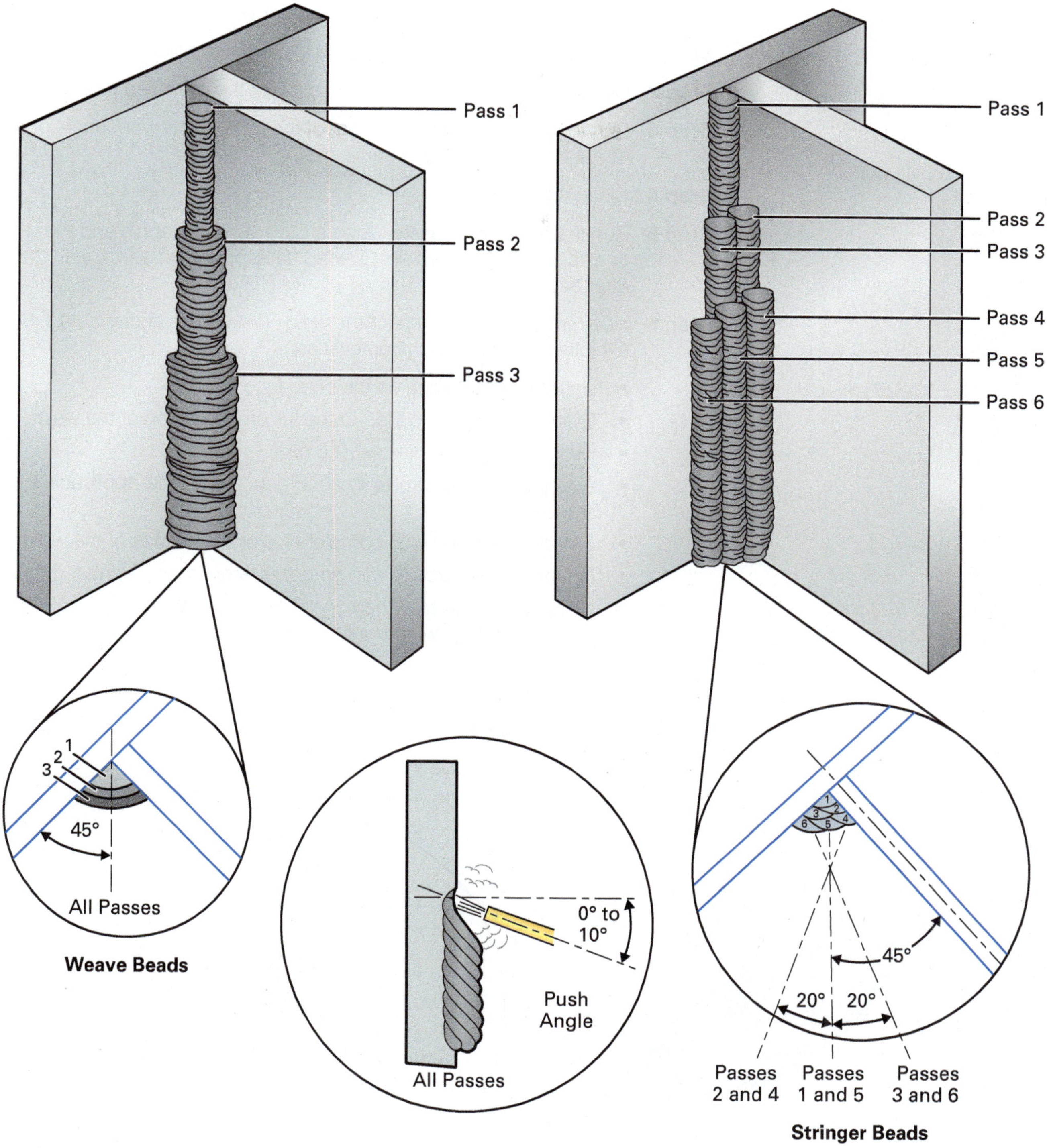

Figure 25 Multipass 3F bead sequences and work angles for weave and stringer beads.

Repeat vertical (3F) fillet welding using stringer beads. Use a slight oscillating motion, and pause slightly at each toe to prevent undercut. For stringer beads, use a 10° to 15° push angle and the required work angles. Refer to *Figure 25* for the bead sequence and work angles. Remember to maintain the 10° to 15° push angle as you weld uphill.

2.4.4 Practicing Overhead (4F) Position Fillet Welds

Practice overhead (4F) fillet welding by welding multipass fillet welds in a T-joint. Use an aluminum filler metal as directed by your instructor. When making overhead fillet welds, pay close attention to the work angle. The work angle is about 45° for the first bead, and then it is adjusted for all other welds.

Follow these steps to make an overhead fillet weld:

Step 1 Tack two plates together to form a T-joint for the fillet weld coupon.

Step 2 Clamp or tack-weld the coupon so it is in the overhead position.

Step 3 Run the first bead along the root of the joint. Use a work angle of about 45° and a 5° to 10° push angle.

Step 4 Clean the weld.

Step 5 Run the remaining passes using a 5° to 10° push angle and the work angles shown in *Figure 26*. Overlap each previous pass. Clean the weld after each pass.

Step 6 Have your instructor inspect the weld. The weld is acceptable if it exhibits the following characteristics:

- Uniform appearance on the bead face
- Craters and restarts filled to the full cross section of the weld
- Uniform weld size of $\pm^1/_{16}$" (1.6 mm)
- Acceptable weld profile in accordance with the applicable code or standard
- Smooth transition with complete fusion at the toes of the weld
- No excessive porosity with no pores larger than $^3/_{32}$" (2.4 mm)
- No undercut greater than $^1/_{32}$" (0.8 mm) deep or 10% of the base metal thickness, whichever is less
- No overlap
- No underfill
- No inclusions
- No cracks

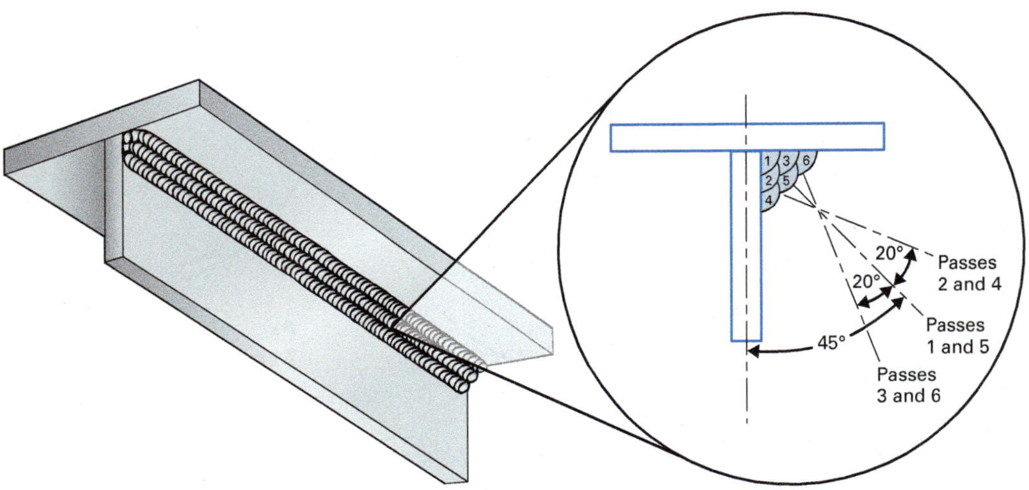

Figure 26 Multipass 4F bead sequence and work angles.

2.5.0 V-Groove Plate Welds

The V-groove weld is a common groove weld made on both plate and pipe. The backing method of welding an aluminum V-groove joint using GMAW is presented in this module. Practicing V-groove welds on plate is good preparation for making the more difficult pipe welds later, which will be covered in another module.

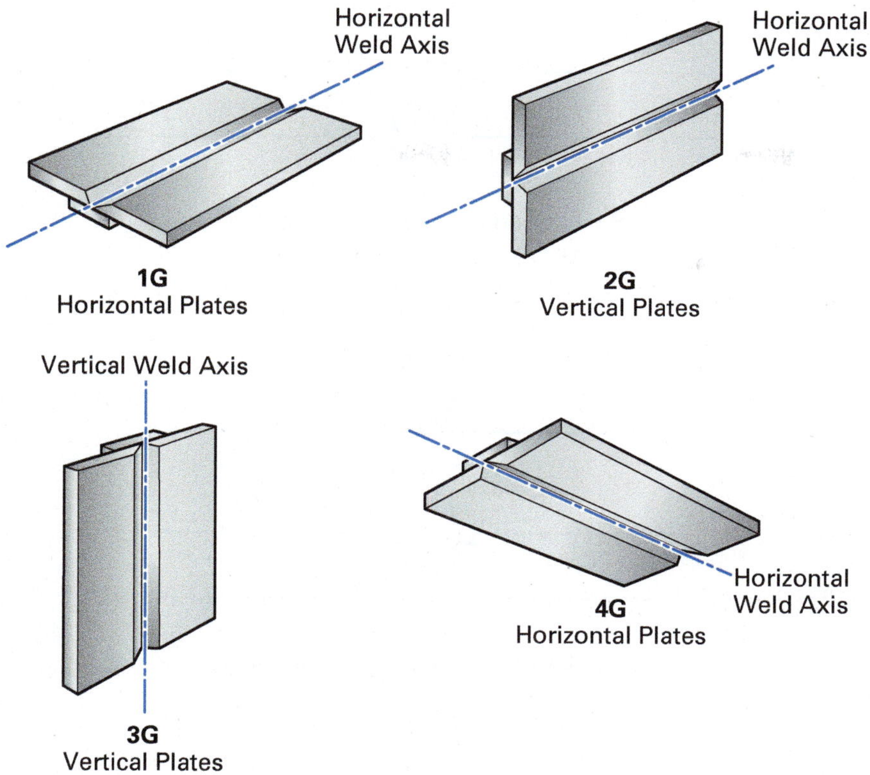

Figure 27 V-groove weld positions.

V-groove welds with backing can be made in all positions. The weld position is determined by the axis of the weld and the orientation of the workpiece. The groove-weld positions are flat (1G), horizontal (2G), vertical (3G), and overhead (4G), as shown in *Figure 27*.

After completing the root pass, it should be cleaned and inspected. Use the proper cleaning method to remove the oxidation that tends to form along the joint. Inspect the root pass to ensure that it exhibits the following characteristics:

- Uniformly smooth face
- Complete fusion
- No excessive buildup
- No undercut greater than $^{1}/_{32}$" (0.8 mm) deep or 10% of the base metal thickness, whichever is less
- No excessive porosity with no pores larger than $^{3}/_{32}$" (2.4 mm)

Practice making GMAW V-groove welds with backing in the 1G, 2G, 3G, and 4G positions. Use the proper size of aluminum filler wire for the passes. Pay special attention to filling the crater at the termination of the weld. Clean each completed bead with a stainless steel wire brush before starting the next bead. Remember that cooling coupons with water can affect the mechanical properties of the base metal. Only practice coupons can be cooled with water. Never cool test coupons with water.

WARNING!

Wear gloves when handling practice coupons and be sure to use pliers when the coupons are hot.

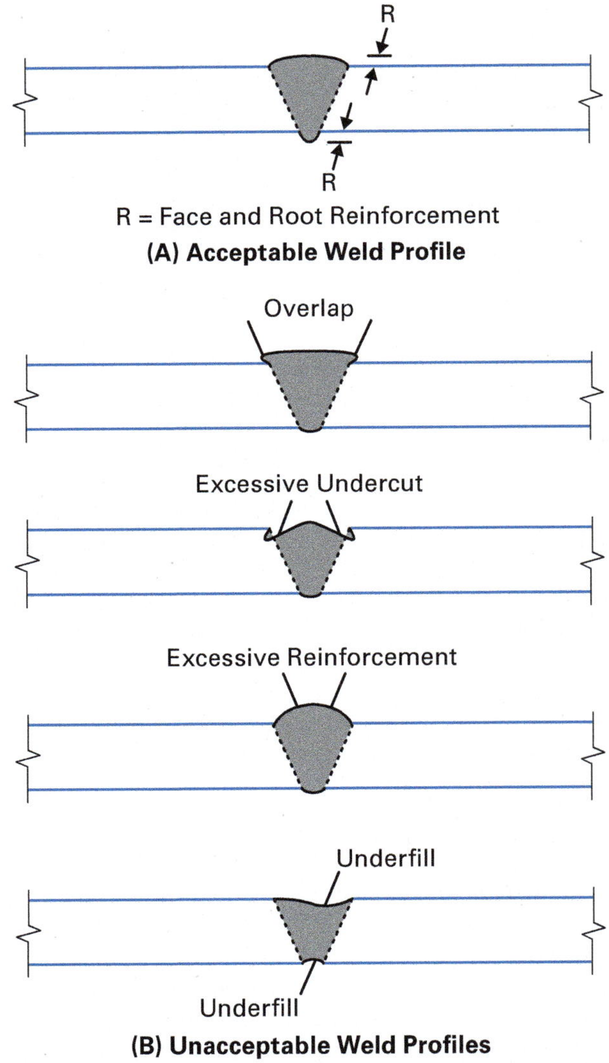

Figure 28 Acceptable and unacceptable V-groove weld profiles.

V-groove welds with backing should be made with slight face reinforcement, not exceeding $^3/_{32}$" (2.4 mm) for materials up to $^3/_8$" (9.6 mm) thick. There should be a gradual transition to the base metal at each toe. Groove welds must not have excessive face reinforcement, any underfill, undercut, or overlap (*Figure 28*). If a groove weld has any of these defects, it must be repaired.

Specific V-Groove Weld Requirements

Refer to your site's WPS for specific requirements on groove welds. The information in this module is provided only as a general guideline. The WPS or site quality standards must be followed for all welds. Check with your supervisor if you are unsure of the specifications for your application.

2.5.1 Practicing Flat (1G) Position V-Groove Welds

Follow these steps to practice making V-groove welds with backing in the flat (1G) position. Refer to *Figure 29* for work angles and the bead sequence.

Step 1 Tack-weld the practice coupon and backing together.

Step 2 Clamp or tack-weld the weld coupon in the flat position above the surface of the welding table.

Step 3 Use a 5° to 10° push angle and a 0° work angle to run the root pass with an appropriate aluminum filler metal.

Step 4 Clean the root pass.

Step 5 Run the remaining passes using a 5° to 10° push angle and the bead sequence and work angles shown in *Figure 29*. Clean the weld between each pass.

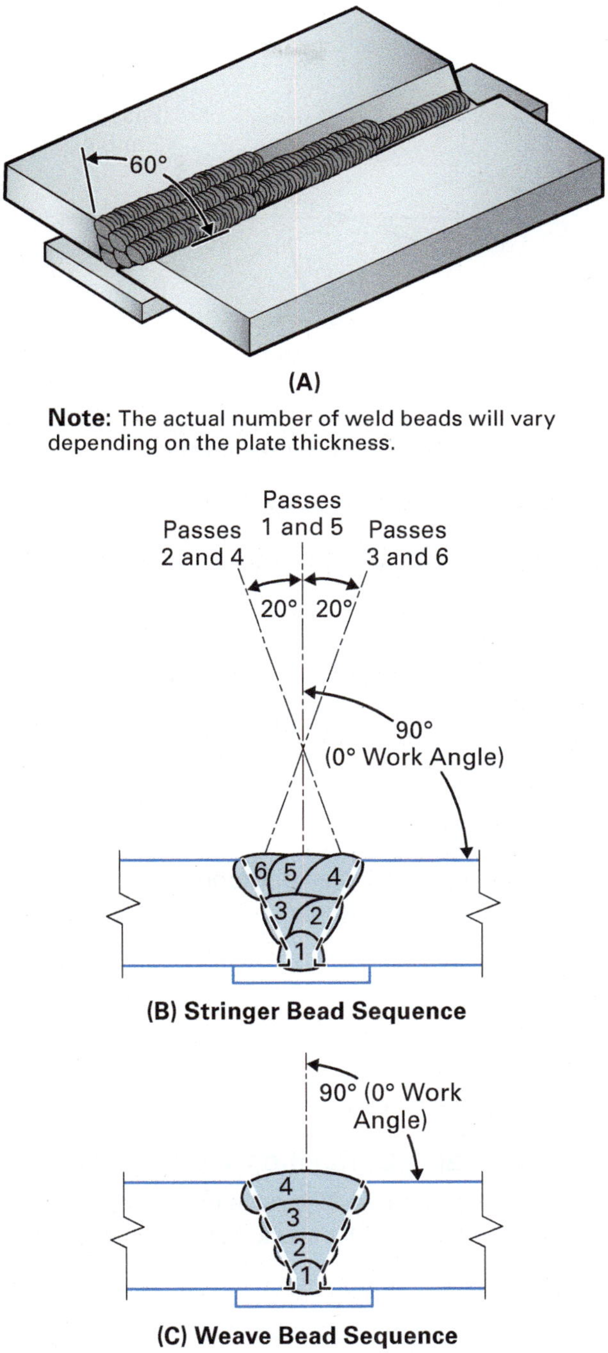

Figure 29 Multipass 1G bead sequence and work angles.

2.5.2 Practicing Horizontal (2G) Position V-Groove Welds

Follow these steps to practice making V-groove welds with backing in the horizontal (2G) position; refer to *Figure 30* for work angles and the bead sequence:

Step 1 Tack-weld the practice coupon and backing together.

Step 2 Clamp or tack-weld the coupon in the horizontal position.

Step 3 Use a 5° to 10° push angle to run the root pass with an appropriate aluminum filler metal.

Step 4 Clean the weld.

Step 5 Run the remaining passes using a 5° to 10° push angle and the bead sequences and work angles shown in *Figure 30*. Clean the weld between each pass.

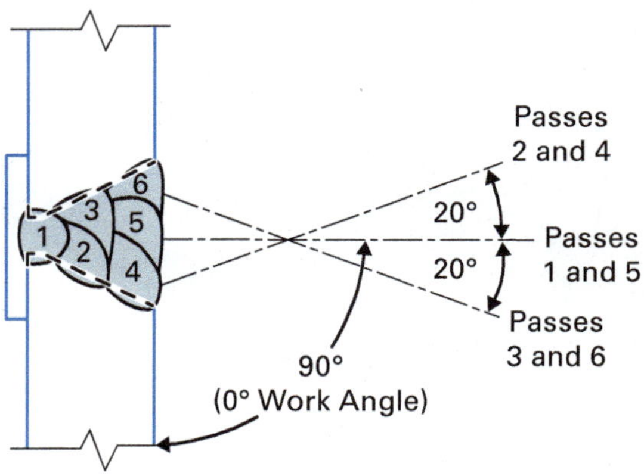

Figure 30 Multipass 2G bead sequence and work angles.

2.5.3 Practicing Vertical (3G) Position V-Groove Welds

Follow these steps to practice making V-groove welds with backing in the vertical (3G) position; refer to *Figure 31* for work angles and the bead sequence. As is true when welding uphill in the 3F position, the push angle is slightly increased in the 3G position. Note that stringer or weave beads may be used for these welds, as specified by your instructor, WPS, or site quality standard.

Step 1 Tack-weld the practice coupon and backing together.

Step 2 Clamp or tack-weld the coupon in the vertical position.

Step 3 Run the root pass uphill using an appropriate aluminum filler metal and a stringer bead. Use a push angle of 10° to 15° and a 0° work angle for the root pass.

Step 4 Clean the weld.

Step 5 Run the remaining passes uphill to complete the weld. Use a 5° to 10° push angle and the bead sequences and work angles shown in *Figure 31* for stringer or weave beads. Clean the weld between each pass.

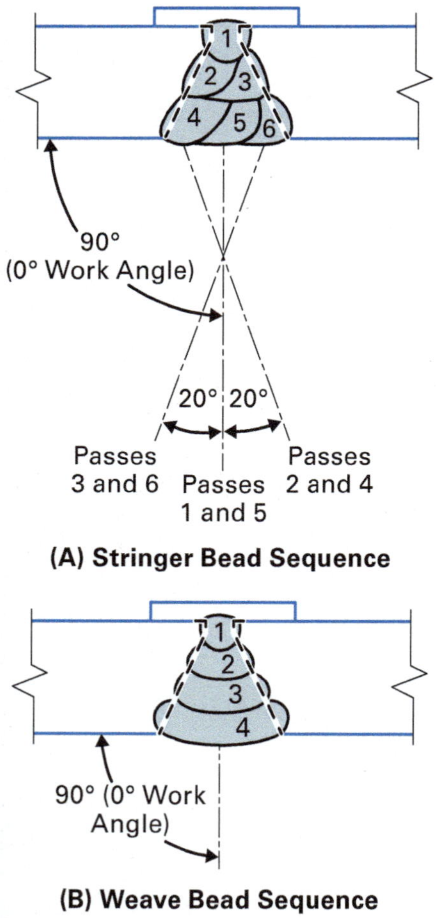

Figure 31 Multipass 3G bead sequences and work angles.

2.5.4 Practicing Overhead (4G) Position V-Groove Welds

Follow these steps to practice making V-groove welds with backing in the overhead (4G) position; refer to *Figure 32* for work angles and the bead sequence:

Step 1 Tack-weld the practice coupon and backing together.

Step 2 Clamp or tack-weld the coupon in the overhead position.

Step 3 Run the root pass using an appropriate aluminum filler metal. Use a 5° to 10° push angle and a 0° work angle for the root pass.

Step 4 Clean the weld.

Step 5 Run the remaining passes using a 5° to 10° push angle and the bead sequence and work angles shown in *Figure 32*. Clean the weld between each pass.

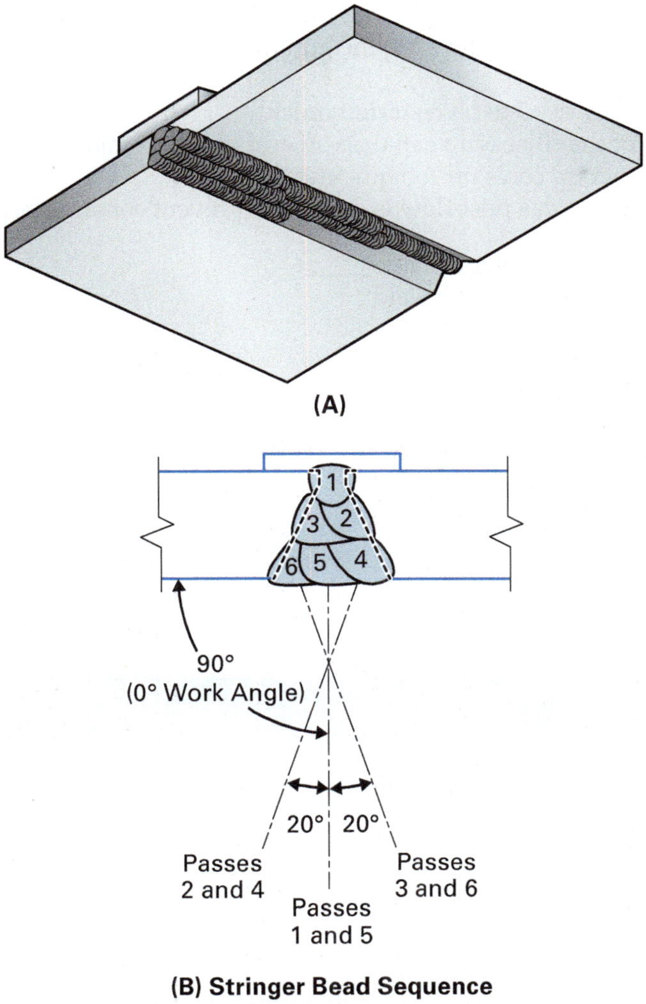

Figure 32 Multipass 4G bead sequence and work angles.

2.0.0 Section Review

1. How much more thermally conductive is aluminum when compared to steel?
   a. Twice as conductive
   b. Three times as conductive
   c. Five times as conductive
   d. Six times as conductive

2. Which of the following is a *correct* statement about aluminum oxide?
   a. Aluminum oxide melts at a much lower temperature than aluminum.
   b. Aluminum oxide is hard enough to use as the grit in grinding wheels.
   c. Aluminum oxide only forms while the aluminum is very hot.
   d. Once aluminum is properly cleaned, aluminum oxide will no longer form.

3. Cracks in aluminum welds that form along the centerline due to expansion and contraction are called _____.
   a. solidification cracks
   b. liquation cracks
   c. oxide cracks
   d. fusion cracks

4. When terminating a weld bead, the welder should allow the weld to solidify before withdrawing the gun from the weld area because the _____.
   a. arc may need to be restarted quickly
   b. tip of the filler wire can drip off and cause an injury
   c. contact tip cools off too quickly
   d. shielding gas post-flow is needed to prevent contamination

5. The 3F position refers to a(n) _____.
   a. horizontal V-groove weld
   b. horizontal fillet weld
   c. vertical fillet weld
   d. overhead V-groove weld

6. The typical work angle used for a GMAW aluminum weld in the 1G position is _____.
   a. 0°
   b. 15°
   c. 30°
   d. 45°

Module 29401 Review Questions

1. OSHA's permissible exposure limit for welding fume from aluminum is set at _____.
   a. 1 mg/m^3
   b. 5 mg/m^3
   c. 10 mg/m^3
   d. 15 mg/m^3

2. In GMAW, the length of wire extending beyond the tip of the welding gun's contact tip is called _____.
   a. setback
   b. stickout
   c. standoff distance
   d. electrode extension

3. Although it is expensive, helium can be added to the shielding gas to _____.
   a. eliminate any chance of porosity
   b. double the welding speed
   c. weld aluminum plate without cleaning it
   d. significantly reduce the welding current

4. The GMAW process for aluminum uses which of the following for the filler metal?
   a. Short lengths of electrode
   b. Tin-coated aluminum rods
   c. A continuous wire electrode
   d. Nickel-coated aluminum rods

5. The aluminum version of rust is known as _____.
   a. aluminum sunburn
   b. aluminum patina
   c. aluminum tarnishing
   d. aluminum oxide

6. The tendency to form weld porosity occurs more in aluminum alloy that contains _____.
   a. nitrogen
   b. magnesium
   c. carbon
   d. hydrogen

7. Cast aluminum has a high silicon content for increased _____.
   a. oxidation resistance
   b. corrosion resistance
   c. fluidity
   d. strength

8. Aluminum that can be formed into plates or sheets, rods, bars, and wire is called _____.
   a. wrought
   b. cast
   c. ingot
   d. fluid

9. Special care should be taken to raise the temperature of cold aluminum slowly because rapid rises in temperature cause _____.
   a. distortion
   b. loss of strength
   c. loss of ductility
   d. condensation

10. The main cause of porosity in aluminum welds is the _____.
    a. absorption of hydrogen into the weld pool
    b. absorption of argon into the weld pool
    c. use of helium in the shielding gas supply
    d. use of an incorrect filler metal

11. When making a weave bead, take care at the toes to _____.
    a. avoid pauses on the edges of the base metal
    b. touch the electrode to the base metal
    c. avoid oscillations on the base metal
    d. ensure proper fusion to the base metal

12. The push angle of an aluminum GMAW weld is increased for which of the following weld positions?
    a. 1F
    b. 2F
    c. 3F
    d. 4G

13. The first bead on an overhead fillet weld is made with a work angle of approximately _____.
    a. 5°
    b. 20°
    c. 25°
    d. 45°

14. The axis of the weld and orientation of the workpiece determine the weld _____.
    a. quality
    b. material
    c. position
    d. strength

15. Due to the potential risk of changes to the mechanical properties of the base metal, cooling welding coupons with water is only appropriate with _____.
    a. pipe weld coupons
    b. test coupons
    c. practice coupons
    d. steel weld coupons

Answers to odd-numbered questions are found in the Review Question Answer Keys at the back of this book.

Answers to Section Review Questions

Answer	Section Reference	Objective
Section 1.0.0		
1. b	1.1.3	1a
2. d	1.2.0	1b
3. c	1.3.1	1c
Section 2.0.0		
1. d	2.1.0	2a
2. b	2.2.1	2b
3. a	2.2.3	2b
4. d	2.3.4	2c
5. c	2.4.0	2d
6. a	2.5.1	2e

User Update

SCAN ME

Did you find an error? Submit a correction by visiting **https://www.nccer.org/olf** or by scanning the QR code using your mobile device.

MODULE 29404

GMAW – Aluminum Pipe

Source: © Miller Electric Mfg. LLC

Objective

Successful completion of this module prepares you to do the following:

1. Explain how to prepare for GMAW of aluminum pipe and describe specific welding techniques.
   a. Identify common welding safety practices.
   b. Describe how to prepare GMAW equipment for aluminum pipe welding.
   c. Explain how to prepare the area and materials for aluminum pipe GMAW.
   d. Describe how to make aluminum V-groove pipe welds using GMAW.

Performance Task

Under supervision, you should be able to do the following:

1. Make GMAW V-groove welds on aluminum pipe with backing in the 2G, 5G, and 6G positions.

Overview

Gas metal arc welding (GMAW) is one of the more common welding processes. GMAW equipment is available in different sizes and types, however, the basic operating principles of GMAW apply to all makes and models of equipment. This module describes how GMAW equipment is used to weld aluminum pipe and presents the techniques used to produce V-groove pipe welds with backing.

NCCER Industry-Recognized Credentials

If you are training through an NCCER-accredited sponsor, you may be eligible for credentials from NCCER. The ID number for this module is 29404. Note that this module may have been used in other NCCER curricula and may apply to other level completions. Contact NCCER at 1.888.622.3720 or go to **www.nccer.org** for more information.

You can also show off your industry-recognized credentials online with NCCER's digital credentials. Transform your knowledge, skills, and achievements into credentials that you can share across social media platforms, send to your network, and add to your resume. For more information, visit **www.nccer.org**.

Digital Resources for Welding

SCAN ME

Scan this code using the camera on your phone or mobile device to view the digital resources related to this craft.

1.0.0 GMAW Aluminum Basics

Performance Task

1. Make GMAW V-groove welds on aluminum pipe with backing in the 2G, 5G, and 6G positions.

Objective

Explain how to prepare for GMAW of aluminum pipe and describe specific welding techniques.

a. Identify common welding safety practices.
b. Describe how to prepare GMAW equipment for aluminum pipe welding.
c. Explain how to prepare the area and materials for aluminum pipe GMAW.
d. Describe how to make aluminum V-groove pipe welds using GMAW.

Gas metal arc welding (GMAW) is an arc welding process that uses a continuous, consumable, solid wire electrode for the filler metal and a shielding gas to protect the weld zone (*Figure 1*). The GMAW process is commonly used to make welds on aluminum, carbon steel, low-alloy steel, and stainless steel. Typical GMAW air-cooled and water-cooled guns are shown in *Figure 2*.

When welding aluminum at current settings of 200 amps or more, a water-cooled gun is preferred to minimize buildup in the gun and contact tip. The contact tips used for aluminum are slightly larger than those used with steel wire to accommodate the expansion of the aluminum wire as it heats up.

CAUTION

Using contact tips designed for steel wire is not acceptable. Due to the expansion rate of the aluminum wire, binding will likely occur as the weld progresses. Only use contact tips designed for aluminum.

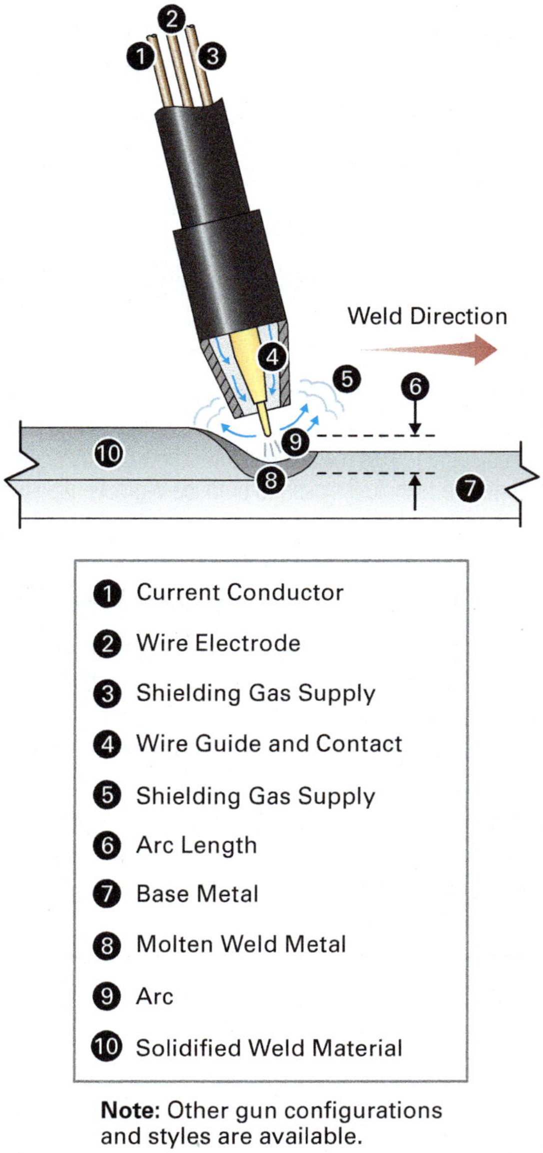

Figure 1 GMAW process.

(A) Inverter Welding Machine

(B) Air-Cooled Gun

(C) Water-Cooled Gun

Figure 2 Inverter-type welding machine with GMAW guns.
Source: Photo courtesy of The Lincoln Electric Company, Cleveland, OH, U.S.A.

GMAW is a fast and effective method for producing high-quality welds. Because this type of welding allows the process to continue for the length of most weld beads, discontinuities and restarts are significantly reduced. Note that it is quite common to use gas tungsten arc welding (GTAW) for the **root pass** and GMAW to complete the **remaining passes**. However, GMAW can be used for root passes as well.

For the educational and training purposes of this module, the dimensions used in the examples are representative of various welding codes and standards and may not be specific to any particular code. When performing welding work, always refer to the Welding Procedure Specification (WPS), applicable welding code, and/or site standard to ensure welds are completed as specified.

Root pass: The initial weld pass made to produce a root bead.

Remaining passes: A nonstandard term used to describe the fill and cover passes as a group.

1.1.0 Safety Practices

Safety considerations are a critical part of preparing for a welding operation. The following is a summary of safety procedures and practices that must be observed while cutting or welding. Before continuing this module, it is strongly recommended that you complete NCCER Module 29101, *Welding Safety*, which provides a thorough review of welding hazards and precautions. Above all, be sure to wear appropriate personal protective equipment (PPE) at all times when welding or cutting.

WARNING!

Welding aluminum is associated with immediate and long-term hazards and health risks. Reflected ultraviolet (UV) light is especially challenging when welding aluminum. Aluminum's highly reflective surface poses a big concern. When welding aluminum, ensure all exposed skin is covered by wearing proper PPE and using welding screens to protect others. Additionally, it is important to follow all other Occupational Safety and Health Administration (OSHA) safety precautions.

Short-term exposure to welding fumes may cause eye and throat irritation, dizziness, and nausea. Prolonged exposure to welding fumes may cause ulcers, damage to the lungs, kidneys, and nervous system, as well as various types of cancer, including lung, larynx, and urinary tract. Due to health risks associated with the welding of aluminum, OSHA has established permissible exposure limits (PELs) for aluminum as well as iron or mild steel. OSHA's PEL limit of 5 mg/m^3 represents the total concentration of fume as an exposure limit over an eight-hour time weighted average. Welders adhere to PELs in addition to following safety practices and protocols of protective clothing, fire prevention, fall prevention, and venting.

1.1.1 Protective Clothing and Equipment

Protective clothing and equipment are a necessary part of every welding task. The following general guidelines review typical PPE required for welding:

- *Welding hood* — A welding hood equipped with a properly shaded lens is best for all forms of welding. The welding hood lens must be an appropriate shade for the application.
- *Clothing and gloves* — Wear proper leather or flame-retardant clothing along with welding gloves for protection from flying sparks, molten metal, and heat.
- *Footwear* — Wear high-top safety shoes or boots. Make sure the tongue and lace area of the footwear is covered by a pant leg. If the tongue and lace area is exposed or if the footwear must be protected from burn marks, wear spats under the pant leg or leather chaps over the front and top of the footwear.
- *Caps and hard hats* — Wear a 100% cotton cap without any mesh material. If a hard hat is required for the environment, use one that allows the attachment of rear deflector material and a face shield. A hard hat with a rear deflector may be required when working overhead and may be required by some employers and jobsites.
- *Ear protection* — Wear earmuffs, or at least earplugs, to prevent sparks from entering the ear canals.

Welders must also familiarize themselves with any unique requirements an employer or job may have regarding PPE.

WARNING!

Welding aluminum is associated with specific hazards and risks that require awareness and safety protocols. Reflected UV light is especially challenging when welding aluminum. Aluminum's highly reflective surface poses a big concern. When welding aluminum, ensure all exposed skin is covered by wearing proper PPE and using welding screens to protect others.

Remember What Grounding Really Means

In the welding trade, the term *ground* is often used to identify the workpiece lead from the welding machine. Welding leads are connected to the positive and negative terminals of the welding machine. The polarity settings of the machine determine the direction of current flow. Neither of the two leads from the welding machine provide any level of actual ground service for the equipment, nor any electrical protection for the welder. True grounding is achieved through a wire lead connected to the machine frame that is then connected to the facility power grounding system.

It is best to use *ground* only as it relates to the safe connection of the welding machine to the power system, providing a path for power to flow to ground should the machine structure become energized. *ANSI Z49.1, Safety in Welding, Cutting, and Allied Processes* requires the worktable to be grounded as well.

1.1.2 Fall Protection and Housekeeping

Welding in the construction industry occurs in a variety of different environments. Welding inside large, open shops exposes a welder to equipment hazards on the floor and overhead. Welding outdoors exposes the welder and the welding job to potential weather hazards. In both environments, welders often must crawl under or reach over objects to complete their work.

At times, welders working indoors and outdoors must work at height. Fall protection equipment is required by OSHA when working at a height of 6' or higher (*Figure 3*).

Fall protection equipment saves lives. Always keep fall protection devices untangled and properly attached. Combined with the other equipment a welder must carry, there is a great deal to manage, but fall protection equipment must always be the priority.

WARNING!

Never attach any kind of welding gear to the safety harness as that could interfere with its ability to perform its intended function.

Regardless of where you work, always keep the work area as neat as possible. This makes the area safer and helps reduce distractions. Mindful actions, such as making sure welding cables are not tripping hazards, contribute to a safer environment. Electrode stubs and unused electrodes must not be left on the floor. They can cause someone to lose their footing and fall. Parts waiting to be welded or left over from previous work are a tripping hazard and should be kept out of the walking area. If the parts have sharp edges, they could cut welding cable insulation or torch gas hoses.

Figure 3 Fall protection for working at height.
Source: Used with permission from MSA

1.1.3 Fire/Explosion Prevention

Flammable gases are commonly found around welding and cutting operations, as are sparks and other significant sources of ignition. For that reason, and because many fire-related incidents over the years have been traced back to welding operations, it is crucial that fire and explosion prevention be given consistent attention. Follow these guidelines to maintain a safe environment:

- Never carry matches, gas-filled lighters, or battery powered devices such as cell phones or vape pens. Sparks can cause the matches or lighters to ignite, or the battery packs to overheat and explode. Any ignition or explosion can cause serious personal injury.
- Never use oxygen to blow dust or dirt from clothing. Traces of the oxygen can remain trapped in the fabric for some time. If a spark hits the fabric during this time, the clothing can burn rapidly and violently out of control.
- Make sure any flammable material in the work area is removed or shielded by a fire-resistant covering. Approved fire extinguishers must be available before attempting any heating, welding, or cutting operations. If a hot work permit and a fire watch are required, be sure these are in place before beginning.
- Ensure all site safety requirements are being observed.
- Never release a large amount of oxygen or use oxygen in place of compressed air. Its presence around flammable materials or sparks can cause rapid and uncontrolled combustion. Keep oxygen away from oil, grease, and other petroleum products.

- Never release a large amount of fuel gas, especially acetylene. Heavier than air, propane tends to concentrate in and along low areas and can ignite at a considerable distance from the release point. Acetylene is lighter than air but is even more dangerous than propane. When mixed with air or oxygen, acetylene will ignite at much lower concentrations than any other fuel gas.
- Even though shielding gases are not toxic, any gas released in a sufficient quantity can displace the oxygen in a small room or enclosed work area, creating an atmosphere with little or no oxygen. Avoid large releases of any type of gas, regardless of its characteristics.
- To prevent fires, maintain a neat and clean work area, and make sure that any metal scrap or slag is cold before disposal.
- Before cutting containers such as tanks or barrels, check to see if they have contained any explosive, hazardous, or flammable materials, including petroleum products, citrus products, or chemicals that decompose into toxic fumes when heated. As a standard practice, always clean and then fill any tanks or barrels with water or purge them with a flow of inert gas, such as nitrogen or argon, to displace any oxygen. Containers must be cleaned by steam cleaning, flushing with water, or washing with detergent until all traces of the material have been removed. Proper procedures for cutting or welding hazardous containers are described in the *American Welding Society (AWS) F4.1, Safe Practices for the Preparation of Containers and Piping for Welding, Cutting, and Allied Processes.*

WARNING!

Welding or cutting must never be performed on drums, barrels, tanks, vessels, or other containers until they have been emptied and cleaned thoroughly, filled with water, or purged with inert gas. All flammable materials and all substances (such as detergents, solvents, greases, tars, or acids) that might produce flammable, toxic, or explosive vapors when heated need to be removed or stable to create a safe atmosphere for welding. Clean containers only in well-ventilated areas. Vapors can accumulate during cleaning, causing explosions and serious personal injury.

1.1.4 Work Area Ventilation

Many gases are introduced to the atmosphere through welding processes. Some of these gases are harmful, but all of these gases can displace or combine with atmospheric oxygen in a confined space. In the case of aluminum, GMAW applies high heat to melt the aluminum and its alloys, which facilitates oxidation. Aluminum oxide particles are formed, which mix with the welding fumes. The welding fumes produce white visible smoke that contain known lung irritants, magnesium oxide and aluminum oxide.

Follow these guidelines, as well as those of your employer, for work area ventilation to prevent personal injuries including long-term respiratory harm:

- Follow all confined space procedures before welding or cutting in a confined space. Use powered fume extraction systems, as needed.
- Never use oxygen to provide ventilation in confined spaces.
- Always perform cutting or welding operations in a well-ventilated area.

The welding of aluminum is associated with specific gases and fumes including ozone, nitrous oxides, carbon monoxide, and carbon dioxide. For example, ozone is a known lung irritant, so proper ventilation needs to be utilized to minimize the risk. Ozone is produced during GMAW of aluminum. The UV rays from the welding arc interact with oxygen in the area, forming ozone gas.

The composition and quantity of the gases and fumes produced from welding aluminum are dependent upon the welding process, filler wire used, coatings on the metal being welded (i.e. paint, anodizing, or Alodine coatings), number of welders, volume of the work area, quality and the amount of ventilation, position of the welder with respect to the fume plume, and presence of contaminants in the atmosphere such as chlorinated hydrocarbon vapors from cleaning and degreasing activities.

For long-term cutting or welding of aluminum and other materials producing toxic gases and fumes, always wear an approved full-face, supplied-air respirator that uses breathing air supplied from outside the work area. For occasional short-term exposure, a HEPA-rated or metal fume filter may be used on a standard respirator.

In addition to controlling work area ventilation for aluminum-specific risks, powered fume extraction equipment must also be used when welding materials that contain zinc, cadmium, hexavalent chromium, manganese, and similar materials that produce hazardous fumes if the area is not naturally well-ventilated. *Figure 4* shows examples of both mobile and portable units.

WARNING!

Breathing in fumes from welding aluminum can cause metal fume fever, also called *welder's flu*. Symptoms include chest tightness, cough, metallic taste, fever, and headache. Proper ventilation on the worksite and individual PPE are necessary to limit exposure and potential risk from fumes created by welding aluminum.

Figure 4 Powered air purifying respirator (PAPR) and localized fume extraction equipment.

Source: Photo courtesy of The Lincoln Electric Company, Cleveland, OH, U.S.A.

Friction Stir Welding

The marriage between steel and aluminum has long been a problem in the welding world due to their extreme differences in thermomechanical properties. Fusion welding of these materials tends to form compounds that are too brittle for reliability. Friction stir welding (FSW) is changing that, and it is equally effective with aluminum alone.

FSW works by applying a tool to the surface of the two metals along the desired joint. The process is autogenous (requires no filler metal). Through friction alone, enough heat is generated to join the two metals, although they never fully melt. The process has already been used in a few specialized applications, such as the front sub-frame of the 2013 Honda Accord and the external fuel tank on the space shuttle. Boeing presently uses the process in aircraft structures and automotive wheel manufacturers are making use of it. The process creates less distortion, doesn't produce a significant heat-affected zone (HAZ), and causes no loss of strength associated with the use of a filler metal. Several videos of the process can be found on the internet.

Source: Image copyright TWI Ltd, reproduced by permission

1.2.0 The Welding Machine

Identify a proper welding machine for GMAW, and follow these steps to set it up for use:

Step 1 Verify that the welding machine can be used for GMAW.

Step 2 Know the location of the primary disconnect.

Step 3 Configure the welding machine for GMAW as directed by your instructor (*Figure 5*), ensuring the proper direct current electrode positive (DCEP) polarity is established. Equip the gun with the correct nozzle for the application. Also equip the gun with the correct liner material and contact tube designed for aluminum wire for the diameter of the wire in use.

Step 4 In accordance with the manufacturer's instructions, configure and load the wire feeder and gun with solid electrode wire of the proper diameter as directed by your instructor.

CAUTION

Using contact tips designed for steel wire is not acceptable. Due to the expansion rate of the aluminum wire, binding will likely occur as the weld progresses. Only use contact tips designed for aluminum.

NOTE

The wire feeder drive rollers should be of the U-groove style for aluminum wire, unlike the V-groove rollers used for carbon steel wire.

Step 5 Connect the proper shielding gas (argon or argon/helium) for the application as specified by the wire electrode manufacturer, WPS, site quality standards, and/or your instructor.

Step 6 Connect the clamp of the workpiece lead to the workpiece.

Step 7 Turn on the welding machine and purge the gun as directed by the gun manufacturer's instructions.

Step 8 Set the initial welding voltage and wire feed speed as recommended by the manufacturer or the WPS for the type and size electrode wire being used.

CAUTION

To prevent the electrode wire from contaminating the welds, always keep the wire clean and dry when it is in storage or in the machine.

Wire Electrode Manufacturer's Recommendations

Always follow the manufacturer's recommendations for using shielding gas and for the initial setting of the welding voltage and wire feed speed parameters. These balanced parameters are critical and are based on the welding position, size, and composition of the solid wire.

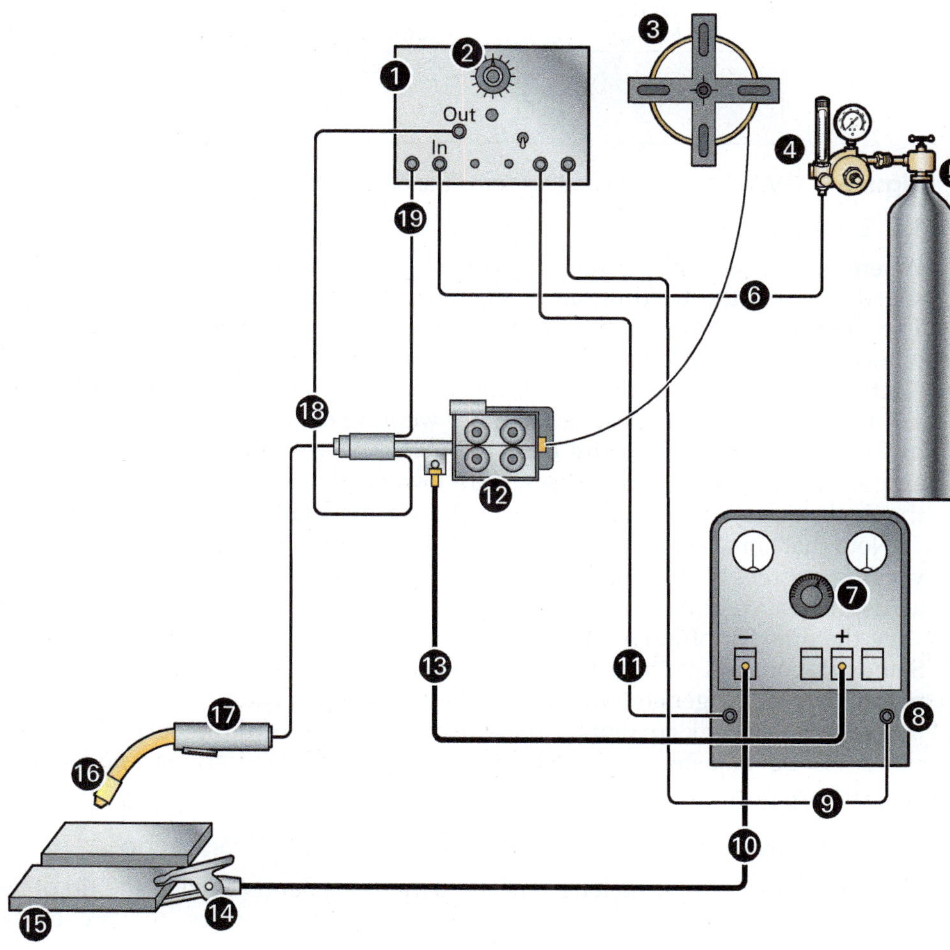

1. Control Unit
2. Wire Feed Speed Control
3. Electrode Wire Reel
4. Regulator/Flowmeter
5. Shielding Gas Source
6. Gas-In Hose
7. Voltage Control
8. Power Source
9. Power to Control Unit
10. Workpiece Lead
11. Contactor Control Cable
12. Wire Feed Drive Assembly
13. Electrode Lead
14. Workpiece Clamp
15. Workpiece
16. Gas Nozzle
17. GMAW Gun
18. Gas-Out Hose
19. Gun Control Cable

Note: The polarity of the gun and workpiece leads is determined by the type of filler metal and application.

Figure 5 Configuration diagram of a typical GMAW machine.

1.2.1 Welding Voltage

Arc length is determined by the voltage, which is set at the power source. Arc length is the distance from the wire electrode tip to the base metal or to the molten pool at the base metal (*Figure 6*). If voltage is set too high, the arc will be too long. This can cause the wire to melt and fuse to the contact tube. Voltage that is set too high can also cause porosity and excessive spatter. Voltage must be increased or decreased as the wire feed speed is increased or decreased. Be sure to set the voltage to maintain consistent spray transfer.

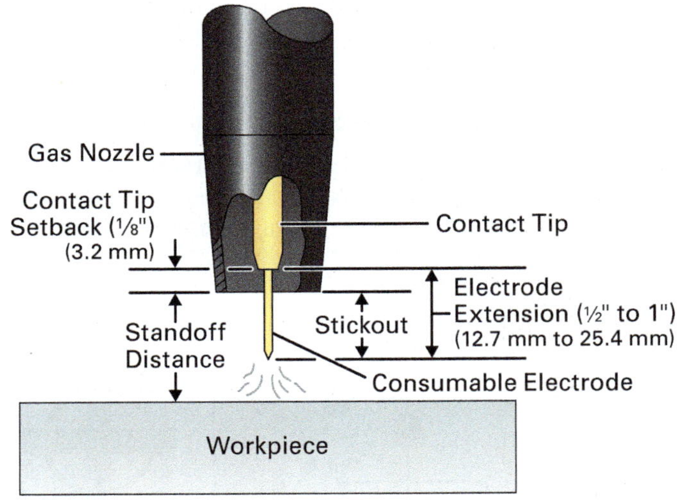

Figure 6 GMAW gun configured for spray transfer.

1.2.2 Welding Amperage

When using a standard constant-voltage welding machine, the electrode feed speed controls the welding amperage after the initial recommended setting. The welding machine provides the amperage needed to melt the wire electrode while maintaining the selected welding voltage. Within limits, when the wire electrode feed speed is increased, the welding amperage and deposition rate are also increased. This results in higher welding heat, deeper penetration, and higher beads. When the wire electrode feed speed is decreased, the welding amperage automatically decreases. With lower welding amperage and less heat, the deposition rate drops, and the weld beads are smaller with less penetration.

Note that some constant-voltage welding machines used for GMAW provide varying degrees of current modification. Examples include slope and induction adjustments. Welding machines with a slope adjustment allow the welder to vary the amount of amperage change in relation to voltage range of the unit. Standard constant-voltage GMAW units have a current slope fixed by the manufacturer for general welding applications and conditions. Pulse transfer units allow the peak current for the pulse and the background current between pulses to be adjusted to match specific welding applications and/or wire electrode requirements.

Travel Speed and Wire Feed Speed

New welders are tempted to turn down the wire feed speed if they have difficulty controlling the weld puddle. Wire electrodes must be run at certain balanced parameters that cannot be changed individually. Voltage, current (if variable), wire feed speed, and travel speed are adjusted and balanced together to control the weld puddle.

1.2.3 Electrode Extension, Stickout, and Standoff Distance

Electrode extension is the length of the wire that extends beyond the tip of the welding gun contact tip. For high-conductivity metal wires, the preheating effect of wire resistance is minimal. Wire speed and voltage settings have a more direct effect on the amount of weld penetration and deposition rate.

Either spray transfer or pulsed spray transfer can be used for aluminum welding. GMAW electrode extension for spray transfer varies from ½" to 1" (12.7 mm to 25.4 mm). *Figure 6* shows the typical electrode extension for the spray transfer GMAW gun configuration. It also shows various gun terms and components. **Stickout** is the distance from the gas nozzle or insulating nozzle to the end of the electrode. **Standoff distance** is the distance from the gas nozzle or insulating nozzle to the workpiece. Contact tip extension or setback is usually dependent on the chosen transfer mode (spray transfer or pulsed spray transfer).

Electrode extension: The length of the wire that extends beyond the tip of the welding gun contact tube.

Stickout: The distance from the gas nozzle or insulating nozzle to the end of the electrode.

Standoff distance: The distance from the gas nozzle or insulating nozzle to the workpiece.

1.2.4 Gas Nozzle Cleaning

As the welding machine is used, weld spatter builds up on the gas nozzle, diffuser, and contact tip. The gas nozzle must be cleaned from time to time with a reamer, round file, or the tang of a standard file. If it is not properly cleaned, it will restrict the shielding gas flow and cause porosity in the weld.

1.3.0 Preparations for Welding

Before welding can begin, the area must be set up, welding equipment must be organized, and the metal to be welded must be prepared. The following sections explain how to prepare for welding.

To practice welding, you will need a welding table, bench, or stand. The welding surface can be steel, but an aluminum surface is preferred. Provisions must be available for placing weld coupons out of position.

To set up the area for welding, follow these steps:

Step 1 Make sure that the area is properly ventilated. If welding fume extraction equipment is not available, make use of doors, windows, and fans.

Step 2 Check the area for fire hazards. Remove any flammable materials before proceeding.

Step 3 Locate the nearest fire extinguisher. Do not proceed unless the extinguisher is charged and you know how to use it.

Step 4 Set up welding screens around the welding area.

1.3.1 Practice Welding Coupons

Because aluminum is easily contaminated, proper coupon preparation is very important. Aluminum pipe coupons must be cut to the desired lengths, cleaned, and beveled to the proper angles before they are positioned for welding. A backing ring must also be cut and cleaned. The pipe can be cut by machining or sawing.

The importance of cleaning aluminum before welding cannot be overstated. There are two related facts about aluminum that help explain why. First, aluminum does not give the welder a visual indication that it is nearing its melting point through color change, as many other common metals do. With no indication that the melting point is approaching, it is easy to apply too much heat. Second, aluminum oxide—the "skin" that forms on the outside—has a melting point roughly three times higher than the aluminum itself. Excessive weld contamination and poor weld quality result when the aluminum base metal is not thoroughly cleaned.

The preferred technique for beveling and/or adding a root face is mechanical machining. Grinding is also acceptable, although machining the grooves and root faces of aluminum pipe is the cleanest and most uniform way to prepare pipe for GMAW. Thermal cutting techniques are not considered acceptable. If thermal cutting techniques must be used to cut aluminum alloys, at least $^1/_8$" (3.2 mm) of the cut surface must be removed by machining or grinding before welding.

WARNING!

Grinding wheels used on aluminum must be made for aluminum only. They should not be used on any other type of metal or on surfaces contaminated with grease or oil. Using other types of grinding wheels on aluminum may cause the grinding wheels to shatter.

CAUTION

Grinding wheels, files, and similar tools used on aluminum should not be used on any other type of metal or on surfaces contaminated with grease or oil.

Pipe weld coupons should be cut from Schedule 40 aluminum pipe that is 6" to 12" (DN150 to DN300) in diameter. Each welded joint requires two coupons of the same size. *Figure 7* shows typical AWS aluminum pipe joint specifications. Depending on the joint configuration and the WPS for a given weld, note that a consumable insert may be required.

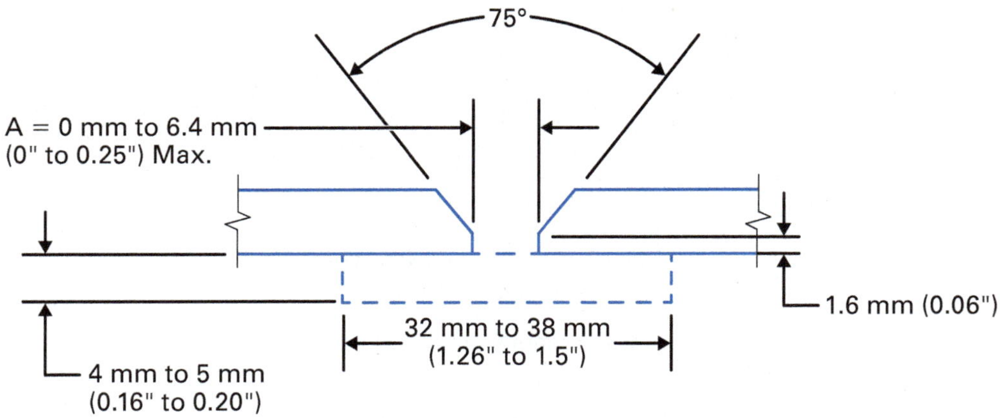

(A) Edge Preparation

Piping Dimensions							Electrode Diameter		Current DCEP	Argon Flow		Number of Passes[a]
Nominal Pipe Diameter Size Number			Outside Diameter (OD)		Wall Thickness							
DN	NPS	Sch.	mm	(in)	mm	(in)	mm	(in)	amp	L/min	(ft³/h)	A = 0
100	4	40	114.3	(4.50)	6.0	(0.24)	1.2	(3/64)	190 – 210	20	(42)	2
125	5	40	141.3	(5.56)	6.6	(0.26)	1.2	(3/64)	205 – 225	20	(42)	2
150	6	40	168.3	(6.63)	7.1	(0.28)	1.2	(3/64)	215 – 235	20	(42)	2
200	8	40	219.1	(8.63)	8.2	(0.32)	1.6	(1/16)	215 – 235	25	(53)	3
250	10	40	273.1	(10.75)	9.3	(0.37)	1.6	(1/16)	215 – 235	25	(53)	3
300	12	40	323.9	(12.75)	10.3	(0.41)	1.6	(1/16)	240 – 260	25	(53)	3

[a] More passes required when A = 6 mm (0.24").

Note: Root opening = 0 for no backing ring or removable backing ring and 6 mm (0.24") for permanent backing ring.

(B) AWS Aluminum Pipe Joint Specifications

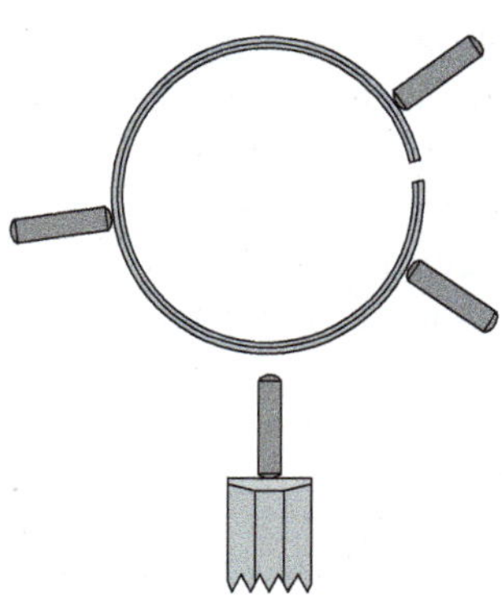

(C) Typical Split Weld Ring with Pegs

Figure 7 AWS aluminum pipe joint specifications.

Source: AWS D10.7M/D10.7:2008, Table 12, reproduced with permission of the American Welding Society (AWS), Miami, Florida (7A-7B)

Preparing Large Pipe

Regardless of the pipe size, there is always a way to cut and bevel it with precision. Clamshell lathes, for example, come in two pieces that are clamped together on the pipe and adjusted to ensure an accurate cut and/or bevel. Clamshell lathes like this one can accommodate pipe sizes up to an astonishing 15' (DN4500). Clamshell lathes are generally hydraulically or pneumatically powered.

Source: Tri Tool Inc.

To prepare pipe coupons for V-groove weld joints with backing, follow these steps:

Step 1 Bevel the end of the pipe and apply a root face as shown in *Figure 7* using any acceptable method discussed in this section.

Step 2 Check the groove and root face. The groove angle should be as seen in *Figure 7*.

Step 3 Cut off a section of the beveled pipe end (4" or 10.2 cm minimum length). For 1G Rotated welding, you may need to cut longer coupons to fit the rollers, if used.

Step 4 Use an approved solvent and a clean rag to remove any grease or oil within a minimum of 1" (25.4 mm) from the weld joint. Clean the inside and outside of the coupons with a grinder or stainless steel brush. Also thoroughly clean the backing ring.

WARNING!

To prevent injury when using cleaning solvents, always comply with the Safety Data Sheet (SDS) guidelines for the solvents in use. Make sure that solvents and wiping materials are removed to a safe distance from welding and grinding operations!

Step 5 Use a permanent backing ring compatible with the size and alloy of the aluminum pipe. Insert the ring inside one of the beveled edges of a pipe coupon section. Make an allowance for the root opening (up to $\frac{1}{4}$" or 6.4 mm) if there are no pins or pegs on the backing ring and center the backing ring at the joint edge. Clamp the ring to the pipe at several places. Ensure that the backing ring is appropriately sized for the pipe and fits snug.

Step 6 Tack-weld the ring to the pipe coupon at four equally spaced positions around the pipe. If the weld ring has three pins instead of four, as shown in *Figure 7*, three tack welds are normally made instead. The tack welds should be narrow and should not exceed 1" (25.4 mm) in length. Clean the tack welds.

NOTE

For this module, the practice pipe coupons will be tacked with a root opening and a permanent backing ring without the use of any backing gas.

CAUTION

If you use a grinding wheel, make sure it is a special high-speed grinding wheel approved for use on aluminum. Make sure that the grinding wheel, and any other cleaning tools such as wire brushes, have not been used on any other metal or used to remove hydrocarbon contaminants.

Step 7 Place the beveled edge of the other pipe coupon section over the backing ring. Adjust for the proper root opening. The coupons must also be aligned so that the resulting welded pipe assembly is straight. Align small-diameter pipe by clamping both pieces to a piece of angle iron. Larger pipe can be aligned with the aid of a pipe alignment jig (*Figure 8*) or by holding a straightedge across the joint, parallel to the **axis** of the pipe. The straightedge must be used on all four quadrants of the pipe in case one or both sections are distorted. Then make the first of four tacks parallel to one of the existing tack welds on the ring.

Axis: The center line running the length of an object. In welding, the term *axis* is most often used in relation to an electrode or a section of pipe.

Step 8 After the first tack weld, check the root opening on the opposite side. Adjust the root opening as necessary and make the second tack weld on the opposite side of the first tack.

Step 9 Check the root opening again and weld the third tack midway between the first two tacks.

Step 10 Apply the fourth tack opposite the third tack and midway between the first and second tacks. There should now be four sets of tack welds evenly spaced (90°) around the pipe coupon as shown in *Figure 9*.

Step 11 Clean and **feather** the tack welds with the edge of a grinding wheel. Feathering the ends of the tack welds with a grinder helps ensure that the tack welds become fused into the root pass.

Feather: To grind the start and/or stop areas of a weld to a tapered edge.

Welding Large-Diameter Pipe

The practice coupons on large-diameter pipe must be greater than 6" (15.2 cm) in length. They need more metal to help absorb the increased heat used to make these welds.

Figure 8 Pipe alignment jig.
Source: Courtesy of Ashley Applegate

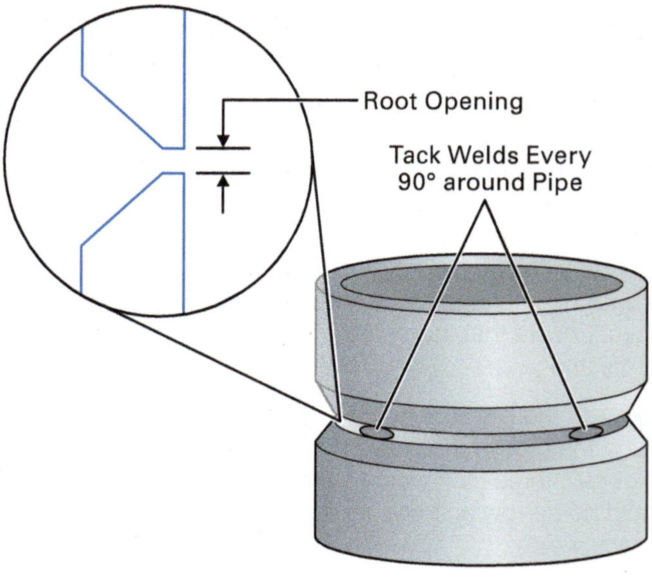

Figure 9 Tacked V-groove weld coupon with backing.

1.4.0 V-Groove Pipe Welds

The V-groove weld with backing is the most common groove weld made on aluminum pipe. These welds are typically used for joining medium- and thick-walled pipe used in various piping systems. Piping containing or carrying material that can cause injury to people or damage the environment, or those that operate at elevated pressures, may be categorized as critical piping. Injury or damage occurs when the piping system or other system components fail to contain the material as designed. Welds in critical piping may be required to meet more stringent requirements. Piping that is used for hydronic heating and air conditioning applications, simple water systems, and similar applications may be assessed using less stringent requirements.

Groove welds may be made in all positions on pipe. The weld position is determined by the **pipe axis**. Four standard weld test positions are used with pipe, as shown in *Figure 10*. The sections that follow explain how to perform these V-groove pipe welds in the four standard positions.

Pipe axis: An imaginary line that runs longitudinally (lengthwise) through the center of a pipe.

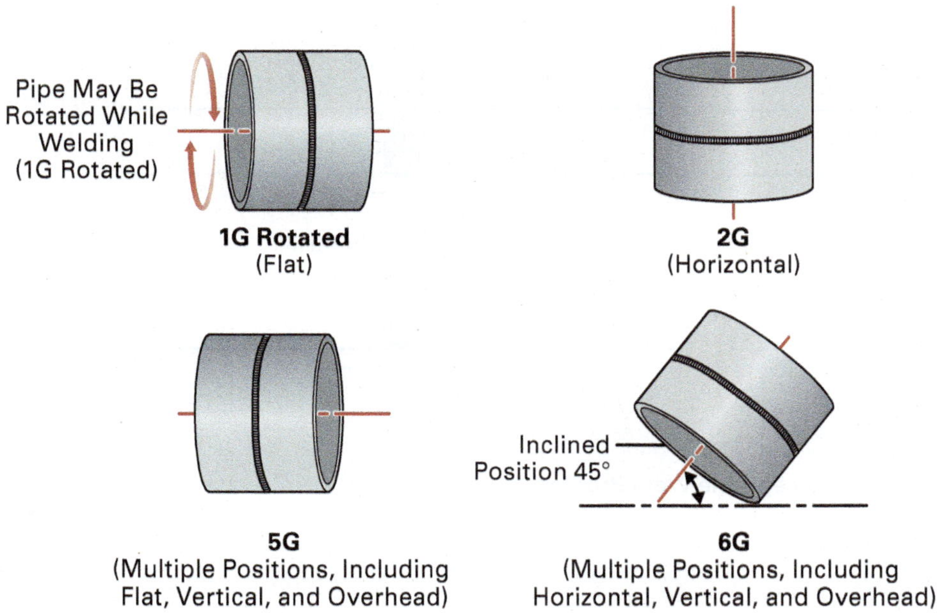

Figure 10 Four basic pipe groove weld test positions.

1.4.1 Acceptable and Unacceptable Pipe Weld Profiles

Pipe groove welds should be made with slight reinforcement and a gradual transition to the base metal at each toe. The root pass should have complete penetration. The face reinforcement can range from being flush to a maximum of $^3/_{32}$" (2.4 mm) for materials up to $^3/_8$" (9.6 mm) thick. Excessively large face reinforcement reduces the pipe's strength.

Completed pipe welds should be inspected to ensure they exhibit the following characteristics:

- Straightness of the bead
- Uniform appearance of the bead face; face and root reinforcement depth in accordance with applicable codes
- Smooth, flat transition with complete fusion at the toes of the weld
- Termination crater filled
- No excessive porosity with no pores larger than $^3/_{32}$" (2.4 mm)
- No undercut greater than $^1/_{32}$" (0.8 mm) deep or 10% of the base metal thickness, whichever is less
- No overlap
- No overfill
- No cracks

If a weld has any of the defects described above and shown in *Figure 11*, it must be repaired.

1.4.2 Weld Travel Speed and Gun Positions

The travel speed must be considered before beginning a weld. Weld travel speed is the speed at which the electrode tip passes across the base metal in the direction of the weld. It is measured in inches per minute (IPM). The metric value for travel speed may be millimeters or centimeters per minute. Travel speed has a great effect on penetration and bead size. Slower travel speeds build higher beads with deeper penetration. Faster travel speeds build smaller beads with less penetration. Ideally, the welding parameters should be adjusted so that the electrode tip is positioned at the leading edge of the weld puddle during travel.

The gun travel angle in relation to the direction of the weld is defined as shown in *Figure 12*. A push angle is used to allow the shielding gas to cover and protect the base metal ahead as the weld advances.

Arc Voltage

A minimum arc voltage is needed to maintain the spray transfer welding process. However, penetration is not directly related to voltage. Penetration will increase with voltage over a limited range, but it will decrease if the voltage is increased above its optimum.

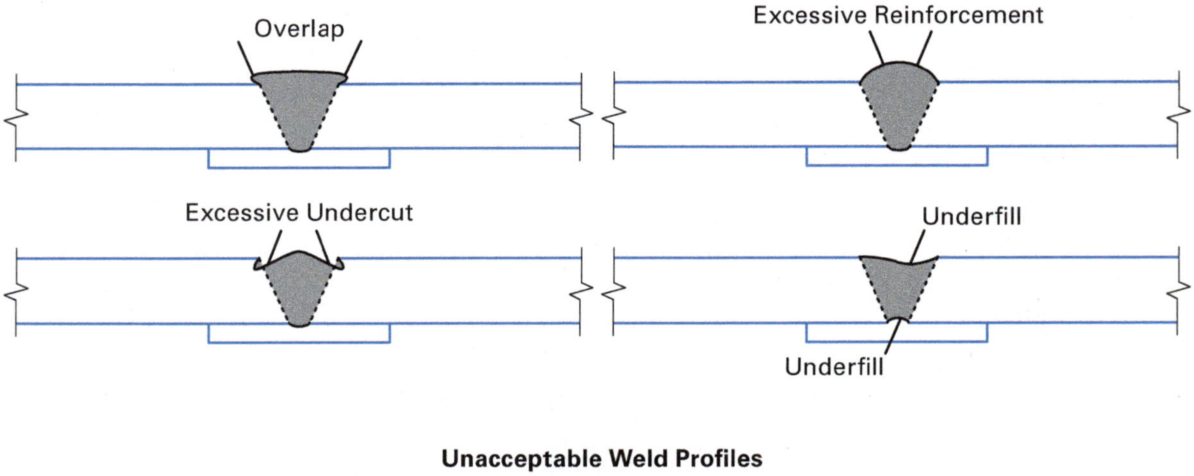

Figure 11 Unacceptable pipe groove weld profiles.

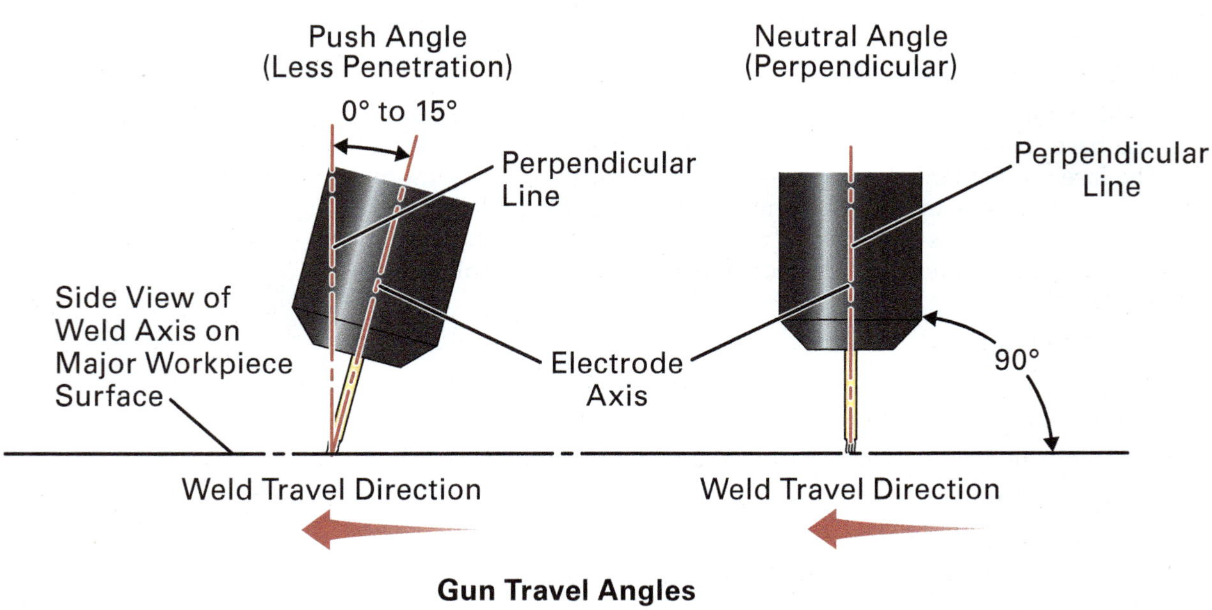

Figure 12 Gun travel and work angles.

1.4.3 Practicing Flat (1G Rotated) Position V-Groove Welds

Practice the 1G Rotated position V-groove pipe welds with backing. Use a shielding gas and aluminum filler wire of a diameter specified by your instructor. Ensure that the contact tip used is the proper size for the aluminum filler wire.

For the root pass, keep the gun work angle at 90° (a 0° work angle) to the pipe axis with a 10° to 15° push angle. Be sure to fuse into the tack welds. Clean the face of the root pass with a wire brush.

When running the remaining passes, use stringer or weave beads with a slight side-to-side motion to ensure fusion at the toes of the weld bead. Try to keep the wire at the leading edge of the weld puddle to ensure proper penetration. Take special care to fill the crater at the termination of the weld. Run all passes at or near the top of the pipe as it is rotated.

Follow these steps to practice V-groove pipe welds with backing in the 1G Rotated position:

Step 1 Prepare and tack-weld the practice pipe weld coupon together as described previously.

Step 2 Position the pipe weld coupon horizontally on two sets of rollers at a comfortable welding height.

Step 3 Make sure that the workpiece clamp is attached directly to the pipe coupon. This will prevent the welding current from passing through the roller bearings or arcing between the rollers and the pipe coupon.

Step 4 Run the root pass using a slight motion. Place a tack weld at the eleven o'clock position. Start the weld bead on the tack weld and advance toward the twelve o'clock position.

Step 5 Roll the pipe as necessary to keep the weld in the flat position.

Step 6 Clean the weld.

Step 7 Use the same rolling procedure to make the remaining passes. Use stringer or weave beads as applicable. Pay particular attention to ensure proper fusion and prevent excess buildup. Overlap the passes and, if using stringer beads, follow the sequence as shown in *Figure 13*. Clean the weld after each pass.

CAUTION

Failure to attach the workpiece clamp to the pipe coupon can result in variations in welding current, damage to the roller bearings, and arcing on the rollers and pipe coupon.

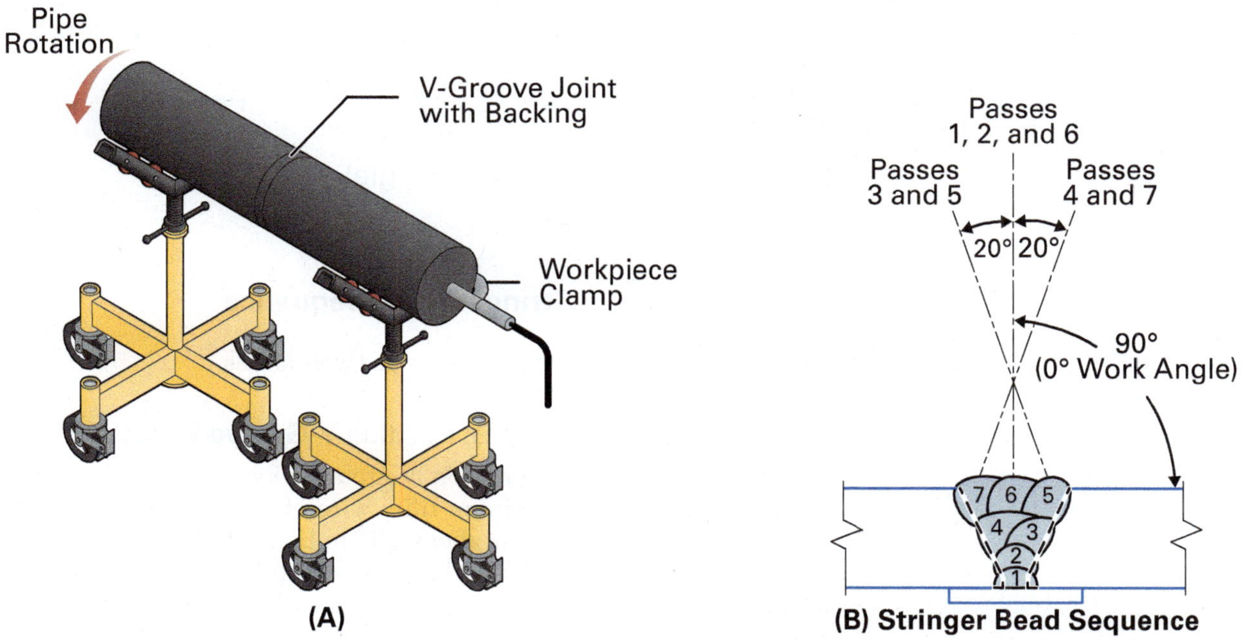

Note: The actual number of weld beads will vary depending on the wall thickness.

Figure 13 Multipass 1G Rotated bead sequence and work angles.

1.4.4 Practicing Horizontal (2G) Position V-Groove Welds

Practice making V-groove pipe welds with backing in the 2G position. Use a shielding gas and an aluminum filler wire of a diameter specified by your instructor.

For the root pass, the gun should be held at a 10° to 15° push angle and at 90° (0° work angle) to the surface of the pipe. To prevent the weld puddle from sagging, the gun work angle can be dropped slightly, but not more than 10°. Be sure to fuse into the tack welds.

When running the remaining passes, use stringer beads and a very slight side-to-side motion to ensure proper fusion at the toes of the weld bead. Try to keep the wire electrode at the leading edge of the weld puddle. Take special care to fill the crater at the termination of the weld. Refer to *Figure 14* for the bead sequence and proper work angles.

Follow these steps to practice V-groove pipe welds with backing in the 2G position:

Step 1 Prepare and tack-weld the practice pipe weld coupon together as described previously.

Step 2 Clamp or tack-weld the pipe coupon with the axis of the pipe vertical.

Step 3 Starting on a tack weld, run the root pass as shown in *Figure 14*. Clean the crater between any necessary restarts. Continue this procedure until the root pass has been completed.

Step 4 Clean the weld with a wire brush.

Step 5 Run the remaining passes at the appropriate work angles. Overlap the passes as shown and clean the weld after each pass.

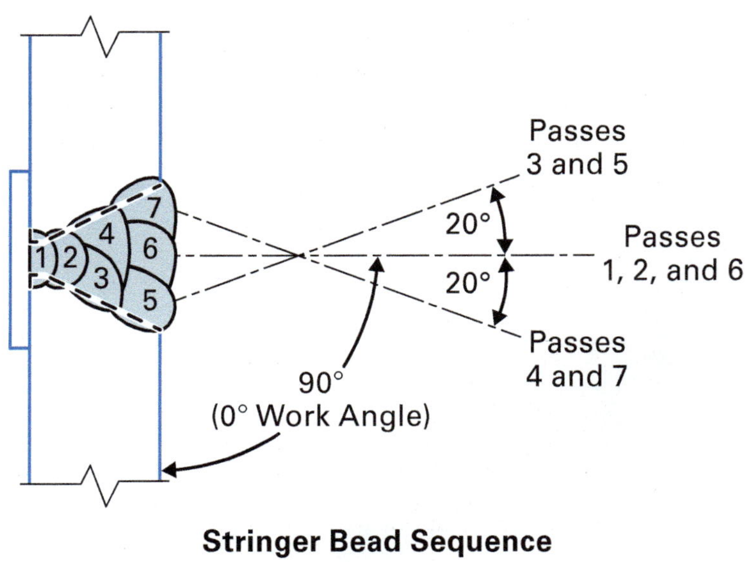

Figure 14 Multipass 2G bead sequence and work angles.

1.4.5 Practicing Multiple (5G) Position V-Groove Welds

Practice the multiple (5G) position V-groove pipe welds with backing. Use a shielding gas and an aluminum filler wire of a diameter specified by your instructor. The gun should be at about a 10° to 15° push angle for the root and all remaining passes. All passes will be applied uphill. *Figure 15* shows the travel positions needed for the passes that are made on both sides of the pipe.

When running the remaining passes as shown in *Figure 16*, use weave beads or stringer beads with a slight side-to-side motion to ensure proper fusion at the toes of the weld. The instructor may direct practice using both techniques. Keep the wire electrode at the leading edge of the weld puddle. Be sure to fill the crater at the termination of the weld.

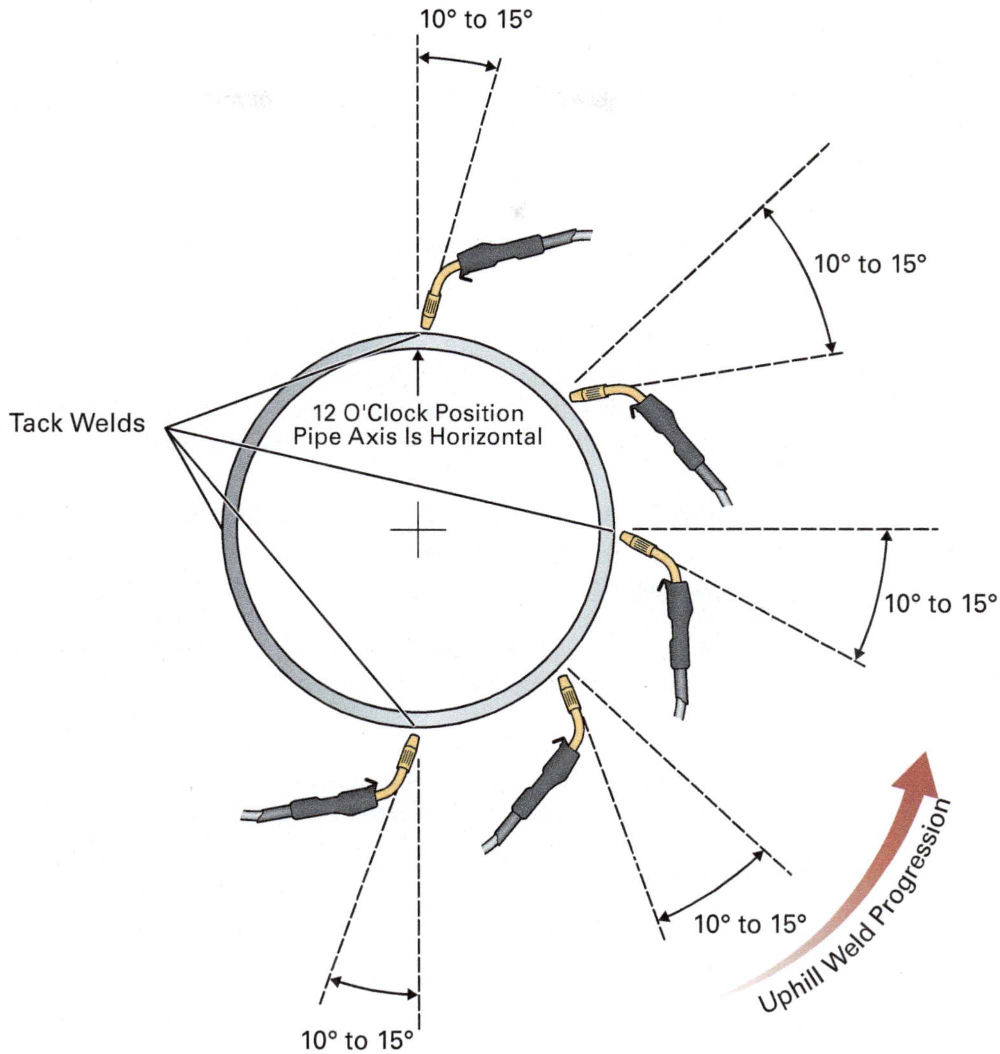

Figure 15 Multipass weld progression for the 5G position.

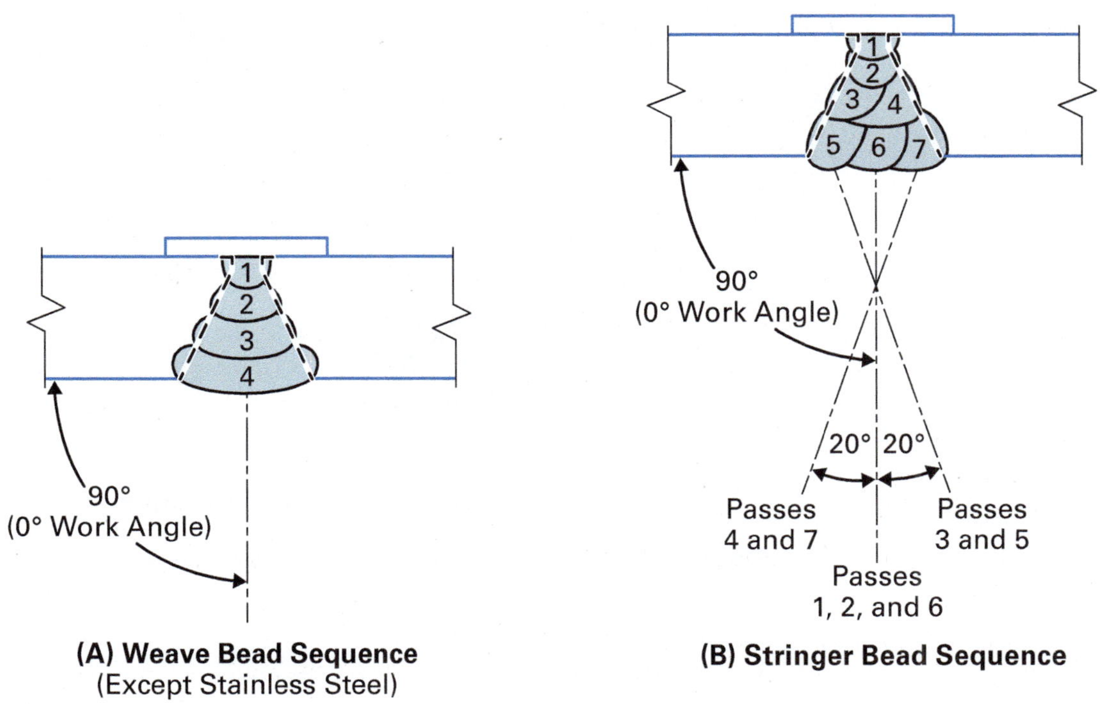

Figure 16 Multipass 5G bead sequences and work angles.

Follow these steps to practice V-groove pipe welds with backing in the 5G position:

Step 1 Prepare and tack-weld the practice pipe weld coupon together as described previously.

Step 2 Clamp or tack-weld the pipe weld coupon into position with the pipe axis horizontal.

Step 3 Run the root pass with a stringer bead, modifying the travel angle as necessary.

Step 4 Clean the weld with a wire brush.

Step 5 Run the remaining passes as stringer or weave beads in an uphill direction at the appropriate work angles. Overlap the passes as shown in *Figure 16* and clean the weld after each pass.

1.4.6 Practicing Multiple Inclined (6G) Position V-Groove Welds

Practice the 6G multiple inclined (45°) position V-groove pipe welds with backing. Use a shielding gas and an aluminum filler wire that has a diameter specified by your instructor. The gun should be at about a 10° to 15° push angle for the root and all remaining passes. All passes will be applied uphill. *Figure 15* shows the travel positions needed for the passes that are made on both sides of the pipe.

When running the remaining passes, use weave beads or stringer beads with a slight side-to-side motion to ensure fusion at the toes of the weld bead. Keep the wire electrode at the leading edge of the weld puddle. Be sure to fill the crater at the termination of the weld. *Figure 17* shows the stringer bead sequence and work angles for the 6G position.

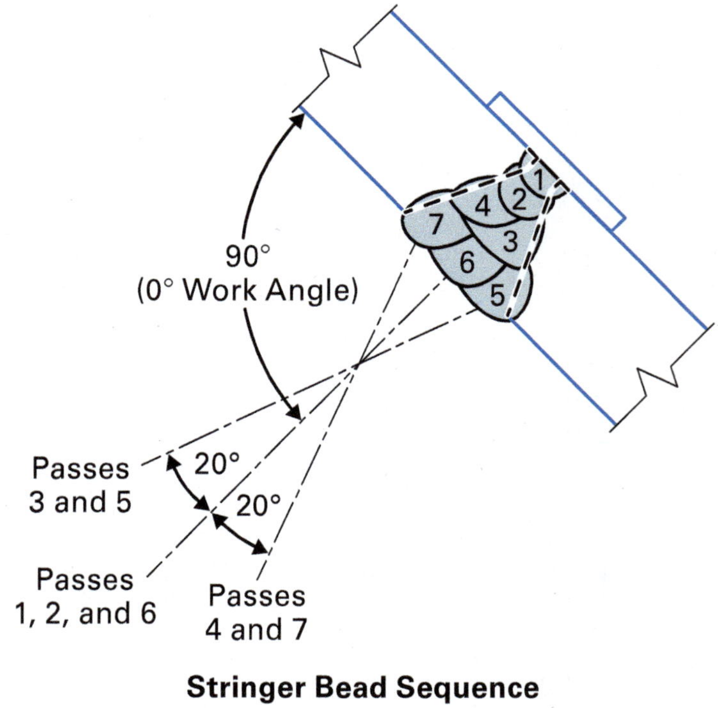

Figure 17 Multipass 6G bead sequence and work angles.

Follow these steps to practice V-groove pipe welds with backing in the 6G position:

Step 1 Prepare and tack-weld the practice pipe weld coupon together as described previously.

Step 2 Clamp or tack-weld the pipe weld coupon into position with the pipe axis inclined 45° to the horizontal plane.

Step 3 Run the root pass with a stringer bead, modifying the travel angle as necessary.

Step 4 Clean the weld with a wire brush.

Step 5 Run the remaining passes in an uphill direction at the appropriate work angles. Overlap the passes and clean the weld after each pass.

NOTE

If desired or required for training purposes, a restricting ring may be added to the 6G position coupon to form a 6GR position coupon.

6GR Test Position

Welding pipe in the 6G position is a common welding test. A ring may be added to test for restricted accessibility (6GR).

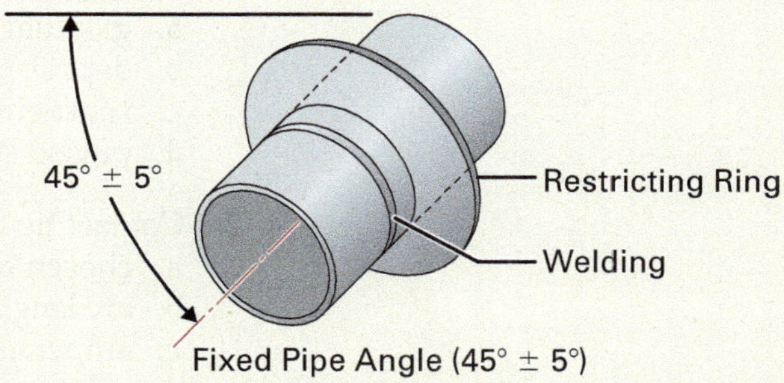

1.0.0 Section Review

1. Caps worn during welding are *best* constructed of _____.
   a. 100% polyester
   b. 100% cotton
   c. a nylon and cotton blend
   d. a nylon and polyester blend

2. When welding amperage is increased, the result is _____.
   a. deeper penetration
   b. less penetration
   c. shorter beads
   d. reduced heat

3. The melting point of aluminum oxide is roughly _____.
   a. one-third that of the aluminum base metal
   b. the same as the aluminum base metal
   c. twice that of the aluminum base metal
   d. three times that of the aluminum base metal

4. When running the remaining passes in the 6G position, you can ensure fusion at the toes of the weld by using a _____.
   a. slight front-to-back motion
   b. slight side-to-side motion
   c. downhill motion
   d. circular motion

Module 29404 Review Questions

1. For the filler metal, the GMAW process for aluminum uses _____.
   a. a continuous, consumable wire electrode
   b. nickel-coated aluminum rods
   c. short lengths of electrode
   d. tin-coated aluminum rods

2. Which statement is *true* about aluminum welding?
   a. All welding gases are harmful.
   b. All welding gases displace or combine with oxygen.
   c. Oxygen is always used to ventilate confined spaces.
   d. Welder's flu is not a concern when welding.

3. The GMAW applications normally used for aluminum are spray transfer and _____.
   a. globular transfer
   b. deposit transfer
   c. short-circuiting transfer
   d. pulsed spray transfer

4. Contact tip extension or setback is usually dependent on the _____.
   a. chosen transfer mode
   b. arc length
   c. amperage
   d. voltage

5. If the weld spatter that accumulates in the gas nozzle is *not* removed, it will restrict the shielding gas flow to the weld and cause _____.
   a. cold fusion
   b. porosity
   c. cracking
   d. worm holes

6. Pipe weld coupons should be cut from Schedule 40 aluminum pipe that has a diameter of _____.
   a. 1" to 3" (DN25 to DN80)
   b. 6" to 12" (DN150 to DN300)
   c. 3" to 18" (DN80 to DN450)
   d. 12" to 16" (DN300 to DN400)

7. When preparing pipe coupons for V-groove weld joints using a backing ring with four pins or pegs, tack-weld the ring to the *first* section of pipe at _____.
   a. two equally spaced positions
   b. four equally spaced positions
   c. six equally spaced positions
   d. eight equally spaced positions

8. Faster travel speeds during GMAW welding of aluminum _____.
   a. prevent termination craters
   b. overheat the contact tip
   c. build taller beads with deeper penetration
   d. build smaller beads with less penetration

9. To ensure proper penetration when welding aluminum pipe, keep the filler metal wire _____.
   a. just above the weld puddle
   b. at the trailing edge of the weld puddle
   c. at the leading edge of the weld puddle
   d. at the center of the weld puddle

10. For GMAW of aluminum pipe, the appropriate push angle for a V-groove weld in the 1G Rotated position is _____.
   a. 15° to 20°
   b. 10° to 15°
   c. 0° to 5°
   d. 0°

Answers to odd-numbered questions are found in the Review Question Answer Keys at the back of this book.

Answers to Section Review Questions

Answer	Section Reference	Objective
Section 1.0.0		
1. b	1.1.1	1a
2. a	1.2.2	1b
3. d	1.3.1	1c
4. b	1.4.6	1d

User Update

SCAN ME

Did you find an error? Submit a correction by visiting **https://www.nccer.org/olf** or by scanning the QR code using your mobile device.

MODULE 29402

GTAW – Aluminum Plate

Source: Photo courtesy of the Lincoln Electric Company, Cleveland, OH, U.S.A.

Objective

Successful completion of this module prepares you to do the following:

1. Explain how to prepare for GTAW aluminum welding and describe specific fillet and groove welding techniques.
   a. Identify common welding safety practices.
   b. Describe common GTAW welding techniques.
   c. Explain how to prepare the area and materials for aluminum GTAW practice.
   d. Explain how to prepare GTAW equipment for welding.
   e. Describe how to accomplish terminations and restarts and create basic GTAW weld beads.
   f. Describe how to make aluminum fillet welds in various positions.
   g. Describe how to make aluminum plate V-groove welds in various positions.

Performance Tasks

Under supervision, you should be able to do the following:

1. Weld a pad on aluminum plate in the flat position using GTAW stringer and weave beads.
2. Make multipass GTAW fillet welds on aluminum plate in the 1F, 2F, 3F, and 4F positions.
3. Make multipass GTAW V-groove welds with backing on aluminum plate in the 1G, 2G, 3G, and 4G positions.

Overview

The gas tungsten arc welding (GTAW) process is unique in that the welder must add the filler metal using the opposite hand. The process is well suited for welding aluminum, using the freehand or the walking-the-cup technique and both stringer and weave beads. This module provides insight into welding aluminum using the GTAW process and an opportunity to hone the necessary skills.

NCCER Industry-Recognized Credentials

If you are training through an NCCER-accredited sponsor, you may be eligible for credentials from NCCER. The ID number for this module is 29402. Note that this module may have been used in other NCCER curricula and may apply to other level completions. Contact NCCER at 1.888.622.3720 or go to **www.nccer.org** for more information.

You can also show off your industry-recognized credentials online with NCCER's digital credentials. Transform your knowledge, skills, and achievements into credentials that you can share across social media platforms, send to your network, and add to your resume. For more information, visit **www.nccer.org**.

Digital Resources for Welding

SCAN ME

Scan this code using the camera on your phone or mobile device to view the digital resources related to this craft.

1.0.0 Aluminum Plate GTAW

Performance Tasks

1. Weld a pad on aluminum plate in the flat position using GTAW stringer and weave beads.
2. Make multipass GTAW fillet welds on aluminum plate in the 1F, 2F, 3F, and 4F positions.
3. Make multipass GTAW V-groove welds with backing on aluminum plate in the 1G, 2G, 3G, and 4G positions.

Objective

Explain how to prepare for GTAW aluminum welding and describe specific fillet and groove welding techniques.

a. Identify common welding safety practices.
b. Describe common GTAW welding techniques.
c. Explain how to prepare the area and materials for aluminum GTAW practice.
d. Explain how to prepare GTAW equipment for welding.
e. Describe how to accomplish terminations and restarts and create basic GTAW weld beads.
f. Describe how to make aluminum fillet welds in various positions.
g. Describe how to make aluminum plate V-groove welds in various positions.

Tungsten: A nonstandard common name for the tungsten electrode used in GTAW; also, the principal metal from which the electrodes are constructed.

Gas tungsten arc welding (GTAW) is an arc welding process used for making high-quality welds on aluminum. This method of welding has also been referred to as *tungsten inert gas (TIG) welding*. The term **tungsten** refers to the tungsten electrode that delivers the arc. *Figure 1* shows the GTAW process and equipment.

This module reviews GTAW equipment safety and setup and explains how to prepare aluminum plate for both fillet welds and V-groove welds in common positions.

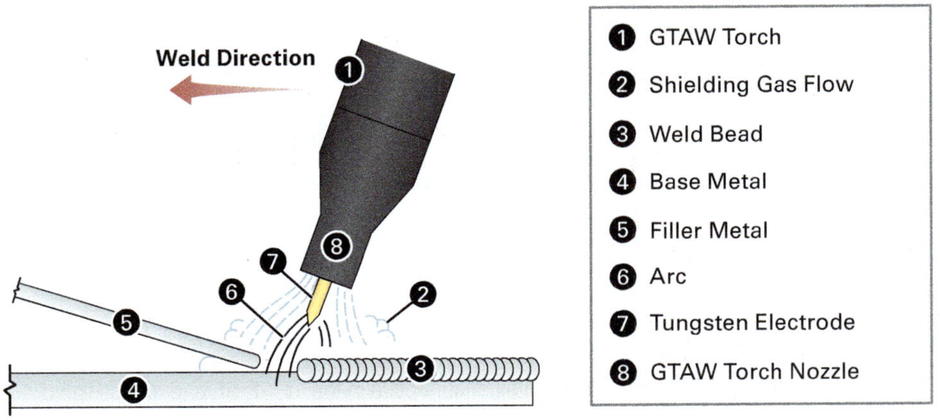

Figure 1 GTAW welding process.

1.1.0 Safety Practices

Safety considerations are a critical part of preparing for a welding operation. The following is a summary of safety procedures and practices that must be observed while cutting or welding. Before continuing this module, it is strongly recommended that you complete NCCER Module 29101, *Welding Safety*, which provides a thorough review of welding hazards and precautions. Above all, be sure to wear appropriate personal protective equipment (PPE) at all times when welding or cutting.

WARNING!

Welding aluminum is associated with immediate and long-term hazards and health risks that can be reduced and controlled, if not eliminated, by following safety practices. Short-term exposure to welding fumes may cause eye and throat irritation, dizziness, and nausea. Prolonged exposure to welding fumes may cause ulcers, damage to the lungs, kidneys, and nervous system, as well as various types of cancer, including lung, larynx, and urinary tract.

Due to health risks associated with the welding of aluminum, the Occupational Safety and Health Administration (OSHA) has established permissible exposure limits (PELs) for aluminum as well as iron or mild steel. OSHA's PEL limit of 5 mg/m^3 represents the total concentration of fume as an exposure limit over an eight-hour time weighted average. Welders adhere to PELs in addition to following safety practices and protocols of protective clothing, fire prevention, fall prevention, and venting.

1.1.1 Protective Clothing and Equipment

Protective clothing and equipment are a necessary part of every welding task (*Figure 2*). The following general guidelines review typical PPE required for welding:

- *Welding hood* — A welding hood equipped with a properly shaded lens is best for all forms of welding. The welding hood lens must be an appropriate shade for the application.
- *Clothing and gloves* — Wear proper leather or flame-retardant clothing along with welding gloves for protection from flying sparks, molten metal, and heat.
- *Footwear* — Wear high-top safety shoes or boots. Make sure the tongue and lace area of the footwear is covered by a pant leg. If the tongue and lace area is exposed or if the footwear must be protected from burn marks, wear spats under the pant leg or leather chaps over the front and top of the footwear.
- *Caps and hard hats* — Wear a 100% cotton cap without any mesh material. If a hard hat is required for the environment, use one that allows the attachment of rear deflector material and a face shield. A hard hat with a rear deflector may be required when working overhead and may be required by some employers and jobsites.
- *Ear protection* — Wear earmuffs, or at least earplugs, to prevent sparks from entering the ear canals.

Welders must also familiarize themselves with any unique requirements an employer or job may have regarding PPE.

WARNING!

Welding aluminum is associated with specific hazards and risks that require awareness and safety protocols. Reflected ultraviolet (UV) light is especially challenging when welding aluminum. Aluminum's highly reflective surface poses a big concern. When welding aluminum, ensure all exposed skin is covered by wearing proper PPE and using welding screens to protect others. Additionally, it is important to follow all other OSHA safety precautions.

Figure 2 GTAW welder with the proper PPE.
Source: © Miller Electric Mfg. LLC

Lightweight Welding Gloves

Soft, flexible, leather gloves are used for light-duty work and for operations involving gas metal arc welding (GMAW), GTAW, brazing, and soldering where more flexibility and dexterity are required. These welding processes produce less spatter than others. More dexterity is required for the welding techniques and for operating torch controls. A wide variety of styles, using various materials, are available. The gloves pictured are used for GTAW and are not as heat-resistant as some other gloves suitable for GMAW. Note, when using welding gloves, always be aware that contamination on gloves can potentially contaminate base metal or filler metal.

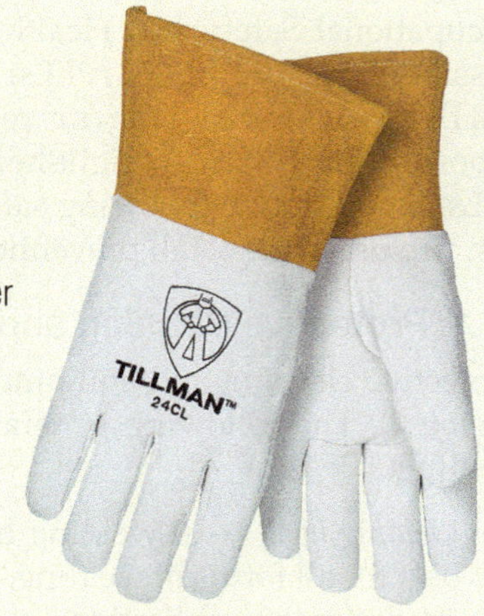

Source: John Tillman Co.

1.1.2 Fall Protection and Housekeeping

Welding in the construction industry occurs in a variety of different environments. Welding inside large, open shops exposes a welder to equipment hazards on the floor and overhead. Welding outdoors exposes the welder and the welding job to potential weather hazards. In both environments, welders often must crawl under or reach over objects to complete their work.

At times, welders working indoors and outdoors must work at height. Fall protection equipment is required by OSHA when working at a height of 6' or higher (*Figure 3*).

Fall protection equipment saves lives. Always keep fall protection devices untangled and properly attached. Combined with the other equipment a welder must carry, there is a great deal to manage, but fall protection equipment must always be the priority. Never attach any kind of welding gear to the harness that could interfere with its ability to perform its intended function.

Regardless of where you work, always keep the work area as neat as possible. This makes the area safer and helps reduce distractions. Mindful actions, such as making sure that welding cables are not tripping hazards, contribute to a safer environment. Electrode stubs and unused electrodes must not be left on the floor. They can cause someone to lose their footing and fall. Parts waiting to be welded or left over from previous work are a tripping hazard and should be kept out of the walking area. If the parts have sharp edges, they could cut welding cable insulation or torch gas hoses.

1.1.3 Fire/Explosion Prevention

Flammable gases are commonly found around welding and cutting operations, as are sparks and other significant sources of ignition. For that reason, and because many fire-related incidents over the years have been traced back to welding operations, it is crucial that fire and explosion prevention be given consistent attention. Follow these guidelines to maintain a safe work environment:

- Never carry matches, gas-filled lighters, or battery powered devices such as cell phones or vape pens. Sparks can cause the matches or lighters to ignite, or the battery packs to overheat and explode. Any ignition or explosion can cause serious personal injury.
- Never use oxygen to blow dust or dirt from clothing. Traces of the oxygen can remain trapped in the fabric for some time. If a spark hits the fabric during this time, the clothing can burn rapidly and violently out of control.

Figure 3 Working at height.
Source: Rich Samanich, M.Ed., SCWI

- Make sure any flammable material in the work area is removed or shielded by a fire-resistant covering. Approved fire extinguishers must be available before attempting any heating, welding, or cutting operations. If a hot work permit and a fire watch are required, be sure these are in place before beginning.
- Ensure all site safety requirements are being observed.
- Never release a large amount of oxygen or use oxygen in place of compressed air. Its presence around flammable materials or sparks can cause rapid and uncontrolled combustion. Keep oxygen away from oil, grease, and other petroleum products.
- Never release a large amount of fuel gas, especially acetylene. Heavier than air, propane tends to concentrate in and along low areas and can ignite at a considerable distance from the release point. Acetylene is lighter than air but is even more dangerous than propane. When mixed with air or oxygen, acetylene will ignite at much lower concentrations than any other fuel gas.
- Even though shielding gases are not toxic, any gas released in a sufficient quantity can displace the oxygen in a small room or enclosed work area, creating an atmosphere with little or no oxygen. Avoid large releases of any type of gas, regardless of its characteristics.
- To prevent fires, maintain a neat and clean work area, and make sure that any metal scrap or slag is cold before disposal.

- Before cutting containers such as tanks or barrels, check to see if they have contained any explosive, hazardous, or flammable materials, including petroleum products, citrus products, or chemicals that decompose into toxic fumes when heated. As a standard practice, always clean and then fill any tanks or barrels with water or purge them with a flow of inert gas, such as nitrogen or argon, to displace any oxygen. Containers must be cleaned by steam cleaning, flushing with water, or washing with detergent until all traces of the material have been removed. Proper procedures for cutting or welding hazardous containers are described in the *American Welding Society (AWS) F4.1, Safe Practices for the Preparation of Containers and Piping for Welding, Cutting, and Allied Processes.*

WARNING!

Welding or cutting must never be performed on drums, barrels, tanks, vessels, or other containers until they have been emptied and cleaned thoroughly, filled with water, or purged with inert gas. All flammable materials and all substances (such as detergents, solvents, greases, tars, or acids) that might produce flammable, toxic, or explosive vapors when heated need to be removed or stable to create a safe atmosphere for welding. Clean containers only in well-ventilated areas. Vapors can accumulate during cleaning, causing explosions and serious personal injury.

1.1.4 Work Area Ventilation

Many gases are introduced to the atmosphere through welding processes. Some of these gases are harmful, but all of these gases can displace or combine with atmospheric oxygen in a confined space. In the case of aluminum, GTAW applies high heat to melt the aluminum and its alloys which facilitates oxidation. Aluminum oxide particles are formed, which mix with the welding fumes. The welding fumes produce white smoke containing the known lung irritants, magnesium oxide and aluminum oxide.

Follow these guidelines, as well as those of your employer, for work area ventilation to prevent personal injuries including long-term respiratory harm:

- Follow all confined space procedures before welding or cutting in a confined space. Use powered fume extraction systems, as needed.
- Never use oxygen to provide ventilation in confined spaces.
- Always perform cutting or welding operations in a well-ventilated area.

The welding of aluminum is associated with specific gases and fumes including ozone, nitrous oxides, carbon monoxide, and carbon dioxide. For example, ozone is a known lung irritant, so proper ventilation needs to be utilized to minimize the risk. Ozone is produced during GTAW of aluminum. The UV rays from the welding arc interact with oxygen in the area forming ozone gas.

The composition and quantity of the gases and fumes produced from welding aluminum are dependent upon the welding process, filler wire used, coatings on the metal being welded (i.e. paint, anodizing, or Alodine coatings), number of welders, volume of the work area, quality and the amount of ventilation, position of the welder with respect to the fume plume, and presence of contaminants in the atmosphere such as chlorinated hydrocarbon vapors from cleaning and degreasing activities.

For long-term cutting or welding of aluminum and other materials producing toxic gases and fumes, always wear an approved full-face, supplied-air respirator that uses breathing air supplied from outside the work area. For occasional short-term exposure, a HEPA-rated or metal fume filter may be used on a standard respirator.

In addition to controlling work area ventilation for aluminum-specific risks, powered fume extraction equipment must be used when welding materials that contain zinc, cadmium, hexavalent chromium, manganese, and similar materials that produce hazardous fumes if the area is not naturally well-ventilated. *Figure 4* shows examples of both mobile and portable units.

WARNING!

Breathing in fumes from welding aluminum can cause metal fume fever, also called *welder's flu*. Symptoms include chest tightness, cough, metallic taste, fever, and headache. Proper ventilation on the worksite and individual PPE are necessary to limit exposure and potential risk from fumes created by welding aluminum.

Figure 4 Powered air purifying respirator (PAPR) and localized fume extraction equipment.
Source: Rich Samanich, M.Ed., SCWI

GTAW in a Bubble

Some metals, such as titanium, molybdenum, and aluminum are more reactive to atmospheric contaminants than others. Problems in the dispersal of the purge gas from the torch, regardless of the reason, represent a larger problem for these metals. The best approach is to complete the weld with the entire workpiece located in an atmosphere of inert gas, preventing any possible contamination from the surrounding air. The purge chamber pictured in this feature provides a suitable environment while also offering portability. The clear, collapsible tent is first evacuated (drawn into a vacuum) to remove the air, and then is filled with a purge gas such as argon from a perforated hose surrounding the base of the tent. The welder uses the openings to reach in and get the job done. Although rigid welding chambers are available, they do not offer the portability of a collapsible product. Complete kits are available as shown in the figure provided, which include a vacuum pump, a flowmeter/regulator, two collapsible tents, and other needed hardware.

(A)

(B)

Source: CK Worldwide, Inc.

1.2.0 GTAW Techniques

GTAW weld bead characteristics and quality are affected by several factors, each of which is influenced by the way the welder handles the torch. These factors include the following:

- Torch travel speed and arc length
- Torch angles
- Torch and filler metal handling techniques
- Consistent feeding of the filler metal

1.2.1 Torch Travel Speed and Arc Length

Torch travel speed and arc length affect the GTAW weld puddle and weld penetration depth of the weld. Slower travel speeds allow more heat to concentrate and form larger, more deeply penetrating puddles. Faster travel speeds prevent heat buildup and form smaller, shallower puddles.

Arc length is the major control for bead width. As the torch is raised and arc length increased, voltage and bead width increase.

1.2.2 Torch Angles

The two basic torch angles that must be controlled when performing GTAW are the work angle and the travel angle. The definition of these angles is the same as for all other types of arc welding.

The torch work angle (*Figure 5*) is a 90° or lesser angle between a line perpendicular to the major workpiece surface at the point of electrode contact and a plane determined by the electrode **axis** and the weld axis. For a T-joint or corner joint, the line is perpendicular to the non-butting member. As shown in *Figure 5*, when the torch is held perpendicular, it is referred to as a *0° work angle*. To properly position beads in a V-groove joint, the work angle should be changed to 20° in either direction.

The torch travel angle is a 90° or lesser angle between the electrode axis and a line perpendicular to the weld axis at the point of electrode contact in a plane determined by the electrode axis and the weld axis. The travel angle used for GTAW is a push angle (*Figure 6*). The push angle is created when the torch is tilted back so that the electrode tip precedes the torch in the direction of the weld. In this position, the electrode tip and shielding gas are being directed ahead of the weld bead being formed. This provides a cleaning action during the positive portion of the alternating current (AC) voltage cycle. Push angles of 15° to 20° are commonly used for aluminum GTAW.

Axis: The center line running the length of an object. In welding, the term *axis* is most often used in relation to an electrode or a section of pipe.

CAUTION

Excessive push angles tend to cause air to be drawn into the weld from under the back edge of the torch nozzle. The air mixes with the shielding gas stream and contaminates the weld.

1.2.3 Torch and Filler Metal Handling Techniques

The freehand technique is the preferred torch and filler metal handling method for performing GTAW on aluminum.

In this method, the torch electrode tip is held just above the weld puddle or base metal (*Figure 7*). The torch is supported by the welder's hand like a pencil. The hand is usually steadied by resting some part of it on or against the base metal. This helps to keep the proper arc length (distance from the electrode tip to the weld puddle). Due to the high thermal conductivity of aluminum, it may help to insert a piece of nonflammable insulation between the gloved hand and the workpiece. As needed, the welder can move the torch tip in a small circular motion within the molten puddle. This maintains the puddle size and advances the puddle. Filler metal is added as needed.

The GTAW filler metal rod is held in the hand that is not supporting the torch. Hold the rod at an angle of 15° to 20° above the base metal surface and in line with the weld. When a 15° to 20° push angle is also used, the rod remains roughly perpendicular to the electrode axis. Always keep the tip of the rod within the shielding gas envelope to protect it from atmospheric contamination and to keep it preheated.

Insert the rod into the leading edge of the weld puddle, using extreme care not to touch the tungsten electrode to the end of the rod. If the tungsten electrode touches the rod or weld puddle, it will become contaminated with filler metal. The electrode must then be removed, the contaminated portion broken off, and the end reformed before welding can continue. To prevent contamination of the electrode by the filler metal, move the electrode to the back edge of the weld puddle as the rod is inserted into the weld puddle's leading edge. Timing those two actions helps prevent tungsten contamination.

1.2.4 Consistent Feeding of the Filler Metal

Feeding the filler metal consistently as the weld progresses is important to ensure the weld puddle develops evenly and produces an even bead. Like all welding techniques, the technique described here requires practice to master but is well worth the effort.

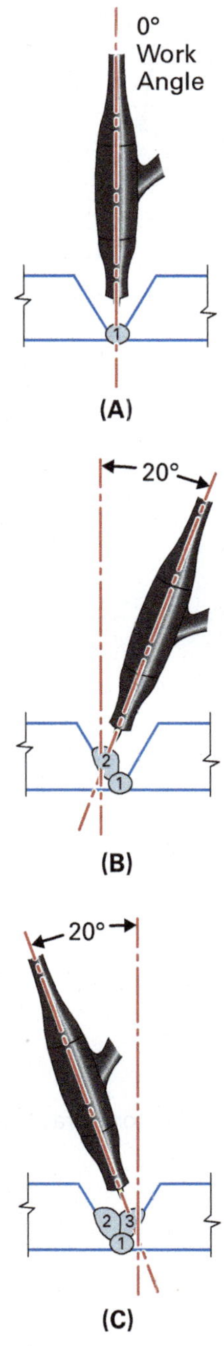

Figure 5 Typical GTAW torch work angles.

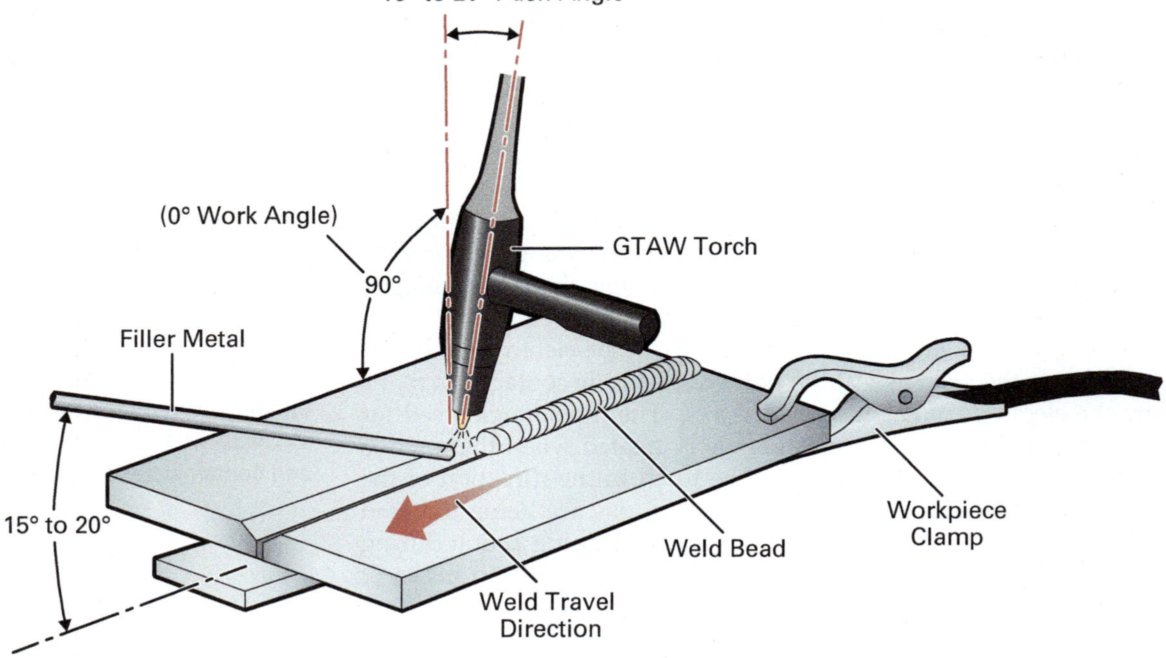

(A) Positions 1G, 2G, and 4G

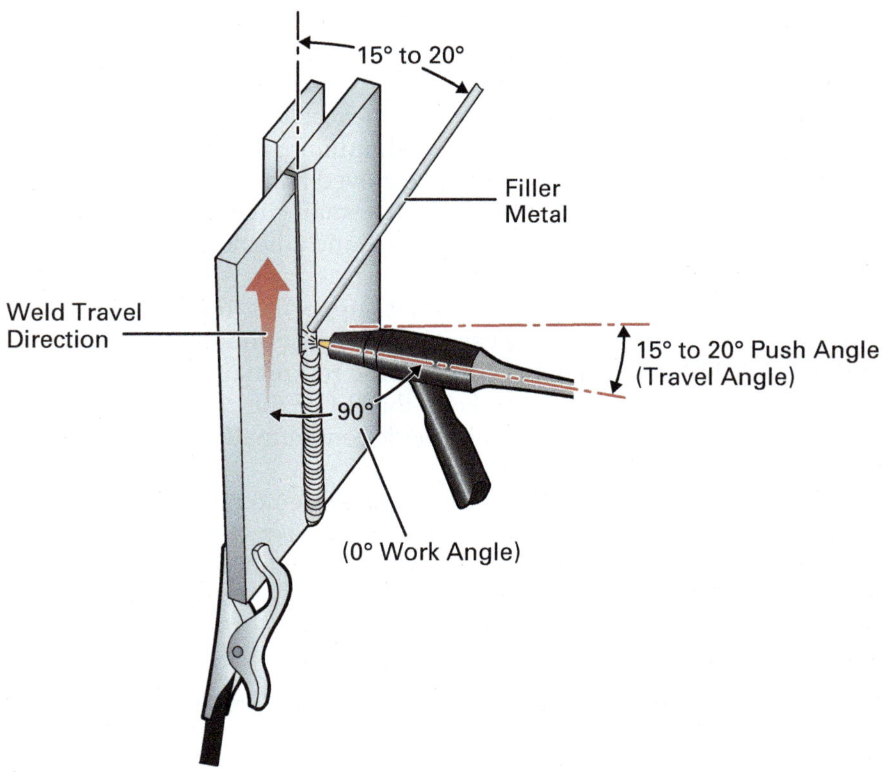

(B) Position 3G (Vertical)

Figure 6 GTAW torch travel angles.

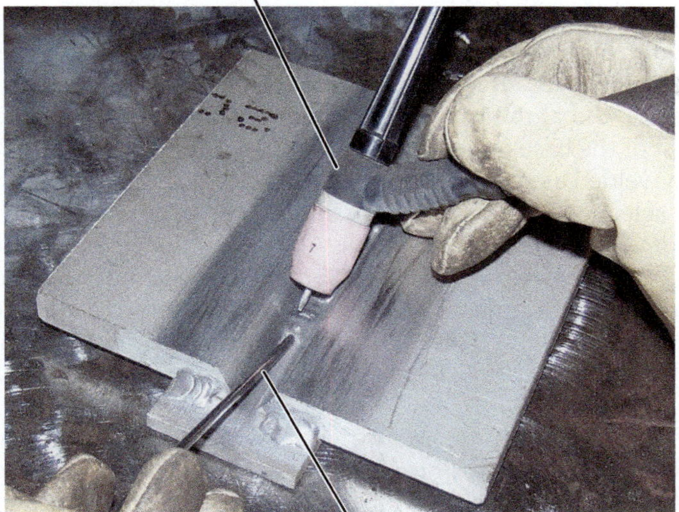

Figure 7 Freehand technique.
Source: Photo courtesy of The Lincoln Electric Company, Cleveland, OH, U.S.A.

Cradle the rod between the pointer finger and the middle finger, similar to the way you might hold a pencil. Then use the thumb to consistently push the rod forward in small motions. Once the thumb is extended pushing the wire forward, quickly bend and reposition the thumb to begin a new forward push. Consistent motion is important, so the thumb must be repositioned quickly to maintain the pace. The thumb can also be used to slowly roll the rod as well during the pushing motion. This technique is easily practiced without actively welding. Simply practice with a rod at home or during free time to become comfortable with the technique before trying it on a weld.

Another technique to consider is referred to as the *finger-pull technique*. The rod is secured between the pointer and middle fingers, which are used to pull the electrode forward. At the end of the forward stroke, the fingers are slipped backward along the rod to reposition, without pulling it backward and out of the weld puddle in the process. Once the two fingers are repositioned, the pulling motion begins again.

1.3.0 Preparations for Welding

Before welding practice can take place, the work area and welding equipment must be prepared, and the coupons to be welded must be cut and cleaned. The following sections explain how to prepare for GTAW of aluminum plate.

To practice welding, you will need a welding table, bench, or stand. The preferred welding surface is aluminum. Provisions must be available for mounting practice welding coupons out of position.

To set up the area for welding, follow these steps:

Step 1 Make sure that the area is properly ventilated. Use doors, windows, fans, and localized fume extraction equipment as needed.

Step 2 Check the area for fire hazards. Remove any flammable materials before proceeding.

Step 3 Locate the nearest fire extinguisher. Do not proceed unless the extinguisher is charged and you know how to use it.

Step 4 Set up welding screens around the welding area.

1.3.1 Selecting Aluminum Filler Metals

The aluminum filler metal typically used for the GTAW process is an uncoated, solid metal filler rod. The filler metal used in aluminum GTAW is chosen based on its metallurgical compatibility with the base metal being welded. Filler metals are available for almost all types of aluminum.

Unlike steel, the aluminum filler metal will not develop a tensile strength equal to that of the base metal. As a result, aluminum weld joints are the weakest link in a welded assembly. This important fact is another reason why proper filler metal selection for aluminum is critical.

Aluminum filler metals are typically sold in cut lengths, most often 36" (91 cm) lengths. Some are available in 18" (46 cm) lengths, but rods can of course be cut to a shorter length if necessary. Common diameters are $^{1}/_{16}$" (1.6 mm), $^{3}/_{32}$" (2.4 mm), $^{1}/_{8}$" (3.2 mm), $^{5}/_{32}$" (4 mm), $^{3}/_{16}$" (4.8 mm), and $^{1}/_{4}$" (6.4 mm). The rods are commonly packaged in 5 lb (2.3 kg), 10 lb (4.5 kg), 25 lb (11.3 kg), or 50 lb (22.7 kg) packages.

All filler rods made in accordance with AWS specifications must have the AWS classification number printed on the shipping container. In addition, filler rod manufacturers frequently print their own unique classification name or number on the shipping container. On some jobs, welders must keep proof of the filler rods used on certain welds. One way to do this is to cut the filler rods in half and burn the rods from their cut ends. The rod ends showing the manufacturers' ID numbers can then be retained for proof that a Welding Procedure Specification (WPS) or other specification has been followed.

Because there are so many types of filler metals available, the welder must be able to distinguish between them and select the correct filler metal for the job. There are two basic factors to consider when selecting aluminum filler metals: the WPS and the base metal composition.

If there is a WPS or site quality standard, the filler rod type will be specified. It will be given as an AWS classification and/or manufacturer's standard. It will sometimes include a manufacturer's name. When a WPS for the weld exists, the filler metal that is specified must be used.

Aluminum filler rods for GTAW are usually selected for their compatibility with the chemical composition of the base metal. There are many distinct types of aluminum base metals and filler metals. *AWS A5.10/A5.10M, Welding Consumables—Wire Electrodes, Wires and Rods for Welding of Aluminum and Aluminum-Alloys—Classification,* provides a guide for the selection of bare aluminum filler metals for specific aluminum base metals and alloys. Additionally, charts and tables provided by filler metal manufacturers assist in selecting the correct filler metal from their product line. An example of a filler metal selection chart from Hobart Brothers Company is provided on the NCCER Digital Resources page, which can be accessed using the QR code at the beginning of this module.

In many cases, only a few filler alloys are used to weld most aluminum alloys. Filler alloys, such as ER5356 and ER4043, can be used for most aluminum welding applications. Remember, however, to always check the WPS or jobsite standards to determine the correct filler metal for the job.

High-quality aluminum welds cannot be achieved unless the filler metal used is clean and high-quality. Contaminated filler metal transfers the contamination to the weld puddle. Filler metal must be used as soon as possible after removal from the shipping or storage packaging. When this is not possible, the filler metal must be stored in airtight storage containers to prevent further oxidation of the rod. Remember that heating aluminum and other metals accelerates the oxidation process dramatically. Once the weld is in progress, the end of the rod becomes hot, resulting in contamination of the rod. For this reason, the tip of the rod should be kept in the shielding gas envelope of the weld to protect it.

Manufacturers' Data for Filler Metals

Most filler metal manufacturers publish paperback welding guides that include information about their filler materials. These booklets can be obtained through your distributor for free or for a small charge. An increasing number of manufacturers also offer apps for filler metal selection that can be used on smartphones, tablets, and similar devices.

CAUTION

Always use filler metal specifications in the WPS or site quality standards, if applicable, to prevent weld discontinuities and satisfy the requirements of the weld.

For all welding exercises associated with this module, use $^3/_{32}$" (2.4 mm), $^1/_8$" (3.2 mm), $^5/_{32}$" (4 mm), or $^3/_{16}$" (4.8 mm) rods, per the aluminum filler material classifications as specified by your instructor. Remove only a small number of filler rods at a time. Store the rest in the package to keep them clean. Before using a filler metal, inspect it for any contamination, such as oxidation, dirt, oil, grease, or burned ends. Contamination will cause defects in the weld. Use emery cloth or stainless steel wool to remove oxides. Use acetone and a clean, oil-free rag to thoroughly clean the filler metal. Snip off any previously burned ends after the filler metal has been cleaned. If the filler metal cannot be cleaned, do not use it. Note that the filler metal and base metal should both be cleaned using the same procedures and cleaning products.

1.3.2 Preparing Aluminum Coupons

Because aluminum is easily contaminated, proper coupon preparation is a priority. Aluminum plate coupons must be cut to the desired lengths, cleaned, and beveled to the proper angles before they are positioned for welding. An aluminum backing must also be cut and cleaned. The basic shapes can be cut by shearing or sawing.

The importance of cleaning aluminum before welding cannot be overstated. There are two related facts about aluminum that help explain why. First, aluminum does not give the welder a visual indication that it is nearing its melting point through color change, as many other common metals do. With no indication that the melting point is approaching, it is easy to apply too much heat. Second, aluminum oxide—the "skin" that forms on the outside—has a melting point roughly three times higher than the aluminum itself. Excessive weld contamination and poor weld quality result when the aluminum base metal is not thoroughly cleaned.

The preferred technique for beveling and/or adding a root face is mechanical machining. Grinding is also acceptable, although machining the grooves and root faces of aluminum joints is the cleanest and most uniform way to prepare aluminum for GTAW. Thermal cutting techniques are not considered acceptable. If thermal cutting techniques must be used to cut aluminum alloys, at least $^1/_8$" (3.2 mm) of the cut surface must be removed by machining or grinding before welding.

CAUTION

Grinding wheels, files, and similar tools used on aluminum should not be used on any other type of metal or on surfaces contaminated with grease or oil.

WARNING!

Grinding wheels used on aluminum must be made for aluminum only. Using other types of grinding wheels on aluminum may cause the grinding wheels to shatter.

Practice coupons are best cut from aluminum plate that is $^1/_4$" (6.4 mm) thick. If this size is not readily available, aluminum plate up to $^3/_8$" (9.6 mm) thick can be used. Material up to $^1/_2$" (12.7 mm) thick can be used with the joint design described herein if welding equipment of sufficient current capacity is available. Material thicknesses greater than $^1/_2$" (12.7 mm) can be used, but a different joint design is required, and welding equipment of extreme current capacity is required when argon gas is used alone. Although it is expensive, helium can be added to the mixture to significantly reduce the welding current required.

NOTE

For welding thin aluminum materials (less than $^1/_4$" or 6.4 mm), there are many other acceptable joint designs that do not require a backing. The joint design used in this module, however, is better suited for practice on the $^1/_4$" (6.4 mm) and thicker material specified above.

Following aluminum cleaning protocols is a crucial step for preventing porosity in aluminum. Porosity in aluminum is caused by hydrogen. Liquid aluminum can dissolve hydrogen. In contrast, solid aluminum dissolves and clears very little hydrogen. A liquid weld puddle will actively pull in any nearby hydrogen. As the weld puddle solidifies, the hydrogen will form porosity. Hydrogen derives from multiple different sources. The first source is hydrocarbons that are found in oils, greases, paint, solvents, and dirt. The second source is water, water vapor, or humidity near the weld.

To minimize porosity, clean the base metal with degreasers or specialized cleaners for aluminum. Use an approved chemical such as acetone to clean oil and grease from the coupons and the metal backing. After oil and grease have been removed, use a stainless steel wire brush to remove all other contaminants and oxidation. Use only light pressure on the brush. Ensure the aluminum is completely dry before welding.

Drastic temperature changes can cause aluminum to form condensation. If working outside or not in a controlled environment, take the aluminum inside. In high humidity locations, controlling or eliminating hydrogen near the weld puddle may be an ongoing challenge. Humidity may come from unexpected sources. For example, welding shops may overlook the use of evaporative coolers, also known as *swamp coolers*, when managing and combating porosity. Even in portions of the country with dry environments and low humidity, evaporative coolers will increase the levels of humidity in welding shops. When critical welds are required in aluminum, look for other methods of controlling the atmospheric conditions in the shop or shielding the weld.

NOTE

Store aluminum following aluminum-specific best practices. Aluminum filler metal should be properly stored and protected from humidity. If aluminum is stored outside, the aluminum should be brought into the shop at least 24 hours prior to welding. Aluminum sheets and plates should be stored vertically; this will help any moisture to drain if the aluminum gets wet. Store aluminum inside whenever possible.

WARNING!

To prevent injury when using cleaning solvents, always comply with the safety guidelines found on the Safety Data Sheet (SDS) for the solvents being used.

Going Green

Conserve Aluminum Practice Coupons

Aluminum is expensive and difficult to obtain for practice purposes. To conserve resources and reduce waste, completely use all weld coupons until all surfaces have been welded upon. Cutting coupons apart and reusing the pieces also conserves materials. Use flat material that cannot be cut into weld coupons to practice running beads. Note that stringer beads can be practiced while leaving enough room between them to practice weave beads as well. Any aluminum material that is no longer suitable for welding practice should be recycled.

The coupons must be shaped to allow the following welds:

- *Stringer beads, weave beads,* and *overlapping beads* — The plate coupons (simple pads) can be any size or shape that can be easily handled.
- *Fillet welds* — Cut the metal into 4" by 5" (102 mm by 127 mm) rectangles for the base and 3" by 5" (76 mm by 127 mm) rectangles for the web (*Figure 8*).
- *V-groove welds with backing* — A properly prepared aluminum V-groove weld joint with backing is illustrated in *Figure 9*. Note that this joint design is not recommended for material thicker than ½" (12.7 mm). As shown in *Figure 10*, cut the metal into 3" by 5" (76 mm by 127 mm) rectangles with one (or both) of the 5" (127 mm) lengths beveled at 30°. Grind a ⅓₂" to ⅛" (0.8 mm to 3.2 mm) root face on the bevel as directed by your instructor. Note that a thicker root face makes it more difficult to achieve proper penetration and consistent fusion. Cut backing strips 1" by 6" (25.4 mm by 152 mm) long from the same alloy as the beveled pieces (*Figure 11*). The backing strip should be sufficiently thick to prevent melt-through.

Coupons and backing strips should be welded within 30 minutes of their preparation to prevent the excessive formation of aluminum oxide. Observe the following steps to prepare each V-groove weld coupon, with backing, for welding:

Step 1 Cut and bevel the coupon components (if required) as shown using any of the acceptable methods discussed in this section. If the bevel has been cut thermally, remember to machine-cut the bevel back roughly ⅛" (3.2 mm) before machining the root face to its proper dimension.

Step 2 Make sure that the surface of the backing strip under the root opening has been cleaned of all oxide coating in the same manner as the coupon. Center the beveled strips on the backing strip with a ⅜" (9.6 mm) root opening as shown in *Figure 11*, and tack-weld them in place. If desired, use a piece of metal or rod with the required thickness as a temporary spacer to set the root opening. Place the tack welds on the reverse side of the joint in the lap formed by the backing strip and the beveled plate. Use three to four ½" (12.7 mm) long tack welds on each beveled plate. Be certain that the backing strip is tight against the beveled plates.

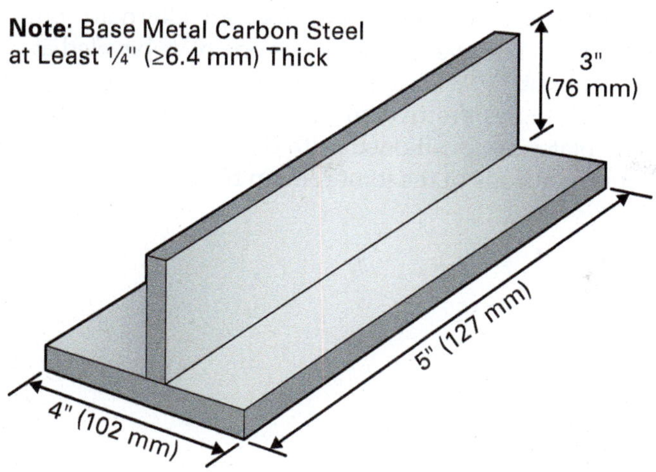

Figure 8 Fillet weld coupons.

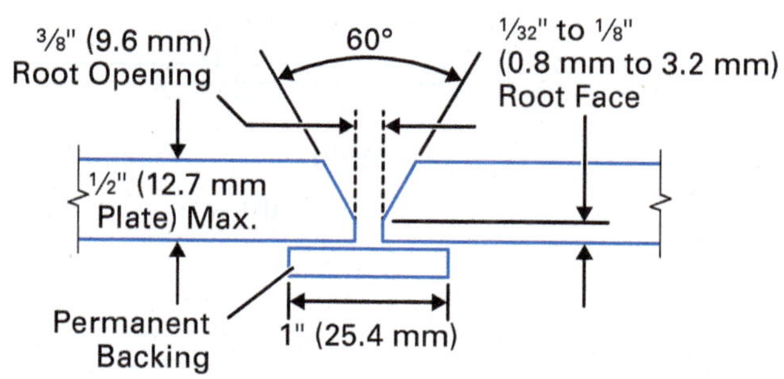

Fit Tolerances	
Root Opening Groove Angle	+¼"; −1/16" (+6.4 mm; −1.6 mm) +10°; −5°

Figure 9 GTAW aluminum weld joint preparation.

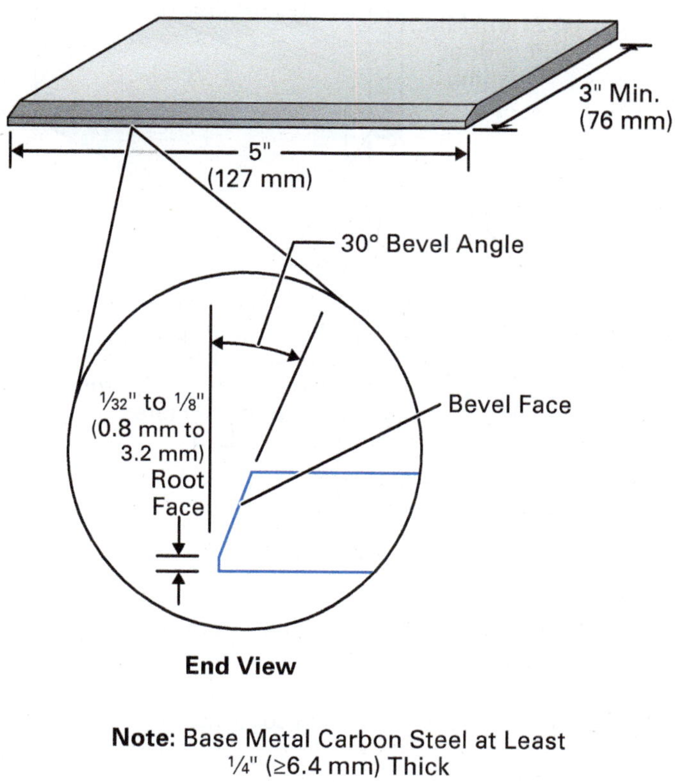

Figure 10 Preparing the V-groove weld coupon.

NOTE

Check with your instructor about whether to use the standard V-groove preparation or the alternate preparation for horizontal weld coupons.

When welding in the horizontal position, an alternate joint preparation (*Figure 12*) can be used. The alternate joint has one plate beveled at about 45° and the other plate at about 15°. As a result, the total groove angle remains at 60°. The standard bevel angles are two 30° angles, which reflect the same groove angle of 60°. The plates are positioned with the 45° beveled plate above the 15° plate. The other dimensions of the joint remain the same.

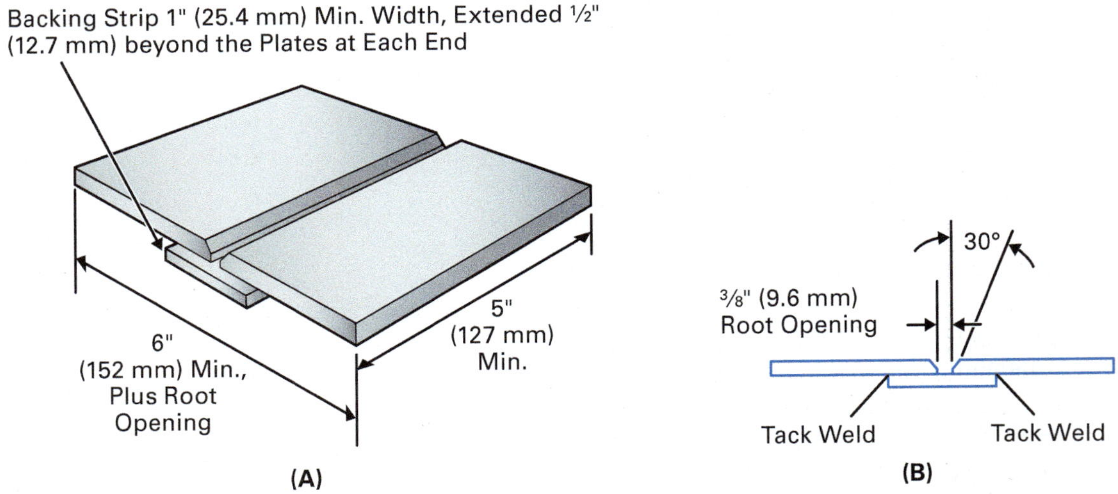

Figure 11 V-groove weld coupon with metal backing.

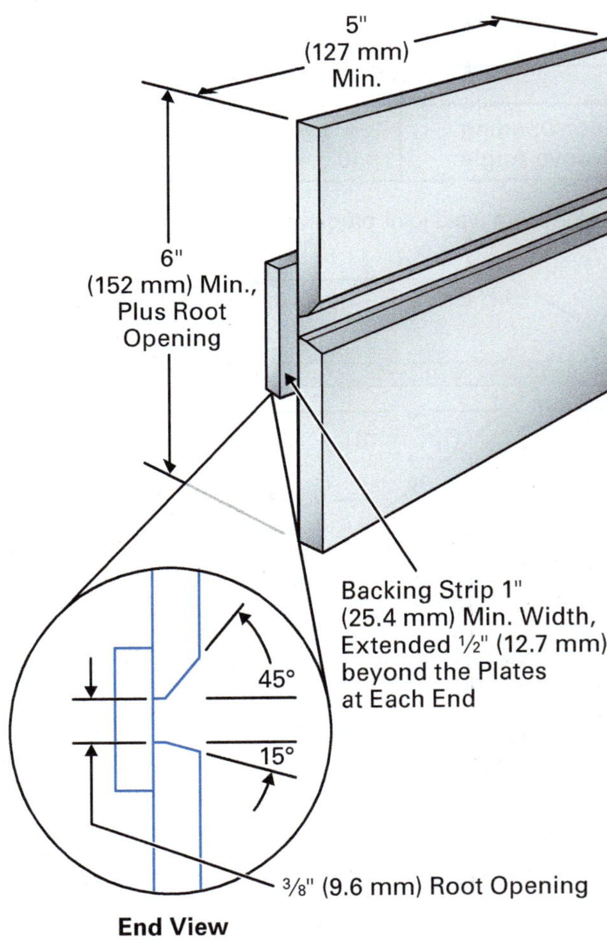

Figure 12 Alternate horizontal aluminum weld coupons.

1.4.0 Welding Equipment

Welding aluminum with GTAW typically requires equipment that provides AC with an argon shielding gas. Aluminum GTAW requires a significant amount of welding current due to its thermal conductivity; heat rapidly flows away from the weld joint. Ensure that the equipment selected is capable of providing sufficient current at a reasonable duty cycle. Welding aluminum at welding current levels less than those required means that the weld cannot progress at the proper speed. As a result, the workpiece will overheat and create an inferior weld. Because of the higher currents associated with aluminum welding, water-cooled torches are recommended. Of course, this will not reduce the temperature of the workpiece itself.

Conventional AC machines used for GTAW are typically based on sine wave or square wave technology. Newer inverter technology units have additional features, such as high-frequency start only and current sequencing (start/stop ramping), that help control puddle formation at the beginning of the weld and in filling the termination crater. Note, scratch start should not be used on aluminum as it can leave tungsten inclusions. The newer inverter technology machines provide additional control over important variables that affect the quality of aluminum welds. The ability to vary both the AC wave balance and frequency is preferred for aluminum welding. Control over the balance between the electrode negative (EN) and electrode positive (EP) portions of the AC sine wave allow the operator to control two important functions:

- As the wave is unbalanced to the EN side, the current-carrying capacity of the electrode is increased and the amount of oxide removal, also called *cleaning action*, is decreased.
- When welding on new material that is properly cleaned, the wave can be set at 80% EN or more. This allows the use of a smaller electrode at higher welding currents.

When welding aluminum with GTAW, a narrow band of the workpiece cleaned by the action of the arc should appear at the toes on each side of the weld bead. *Figure 13* shows examples of stringer beads applied with various EN settings. Note that the plate was not cleaned at all prior to the test to provide a better view of the cleaning action. The lower bead reflects a 95% EN setting that provided insufficient oxide removal. The center bead was applied at an 85% EN setting that provided sufficient cleaning. Finally, the upper bead was the result of a 60% EN setting that produced sufficient oxide removal. Again, properly cleaned aluminum does not need to rely heavily on the arc for oxide removal, allowing for a higher EN setting that provides other advantages.

Figure 13 Cleaning action resulting from various EN settings.
Source: Rich Samanich, M.Ed., SCWI

Truncated: Having the pointed top (vertex) of a conical shape cut off, leaving only a plane section parallel to the base of the cone.

Conventional welding machines in the United States use 60-cycle electrical power as supplied by the line current. (Many other countries use 50 cycles.) Newer inverter machines are able to vary the frequency, potentially increasing it to 300 cycles per second or more. This eliminates the need to use balled pure- or zirconiated-tungsten electrodes and permits the use of a **truncated** EWTh2 or other tungsten electrodes. Increased cycles per second also provides better arc control and allows for smaller weld sizes. Frequencies of 120 cycles to 140 cycles are typically used to weld aluminum.

Although pure tungsten (EWP) electrodes are used for welding aluminum with AC and high-frequency machines, zirconiated-tungsten electrodes (EWZr) are often preferred due to their higher current-carrying capacity and lower susceptibility to contamination. They also hold a balled tip better when welding aluminum.

Table 1 shows the current ranges for various electrode types and sizes.

TABLE 1 Typical Current Ranges for Tungsten Electrodes

Electrode Diameter	Direct Current (Amps)		Alternating Current (Amps)			
	DCEN	DCEP	Unbalanced Wave		Balanced Wave	
	EWCe-2 EWLa-1 EWLa-1.5 EWLa-2 EWP EWTh-1 EWTh-2	EWCe-2 EWLa-1 EWLa-1.5 EWLa-2 EWP EWTh-1 EWTh-2	EWP	EWCe-2 EWLa-1 EWLa-1.5 EWLa-2 EWTh-1 EWTh-2 EWZr	EWP	EWCe-2 EWLa-1 EWLa-1.5 EWLa-2 EWTh-1 EWTh-2 EWZr
0.01" (0.3 mm)	up to 15	*	up to 15	up to 15	up to 15	up to 15
0.02" (0.5 mm)	5 to 29	*	5 to 15	5 to 20	10 to 20	5 to 20
0.04" (1 mm)	15 to 80	*	10 to 60	15 to 80	20 to 30	20 to 60
1/16" (1.6 mm)	70 to 150	10 to 20	50 to 100	70 to 150	30 to 80	60 to 120
3/32" (2.4 mm)	150 to 250	15 to 30	100 to 160	140 to 235	60 to 130	100 to 180
1/8" (3.2 mm)	250 to 400	25 to 40	150 to 210	225 to 325	100 to 180	160 to 250
5/32" (4 mm)	400 to 500	40 to 55	200 to 275	300 to 400	160 to 240	200 to 320
3/16" (4.8 mm)	500 to 750	55 to 80	250 to 350	400 to 500	190 to 300	290 to 390
1/4" (6.4 mm)	750 to 1000	80 to 125	325 to 450	500 to 630	250 to 400	340 to 525

DC and AC Electrode Legend

EWCe-2: Tungsten + 2% Ceria EWLa-1: Tungsten + 1% Lanthana EWLa-1.5: Tungsten + 1.5% Lanthana EWLa-2: Tungsten + 2% Lanthana	EWP: Pure Tungsten EWTh-1: Tungsten + 1% Thoria EWTh-2: Tungsten + 2% Thoria EWZr: Tungsten + Zirconia

Notes:
All values are based on the use of argon as the shielding gas. Other current values may be employed depending on the shielding gas, type of equipment, and application.
*These combinations are not commonly used.

Source: Courtesy of The Lincoln Electric Company, Cleveland, OH, U.S.A.

Once an appropriate welding machine has been selected, follow these steps to prepare it:

Step 1 Verify that the welding machine can be used for GTAW. If a water-cooled torch will be used, select an appropriate cooling unit.

Step 2 Select an air-cooled or water-cooled GTAW torch. Make sure that it is compatible with the welding machine and any optional cooling unit. Water-cooled torches are preferred. Equip the torch with an appropriate cup size for the technique to be used (freehand recommended).

Step 3 Know the location of the primary power disconnect switch.

Step 4 Configure the welding machine for aluminum GTAW welding (*Figure 14*) as directed by your instructor. Configure the torch polarity to AC. Equip the torch with a properly prepared $^{3}/_{32}$" or $^{1}/_{8}$" (2.4 mm or 3.2 mm) EWP or EWZr tungsten electrode.

Step 5 Connect the proper argon or helium-argon shielding gas for the application as directed by your instructor, filler metal manufacturer, WPS, or site quality standards.

Step 6 Connect the clamp of the workpiece lead to the workpiece.

Step 7 Turn on the welding machine. Adjust the flowmeter and purge the torch according to the manufacturer's instructions.

Step 8 Set the initial welding current for the application.

NOTE

Thoriated tungsten electrodes should not be used to weld aluminum. These electrodes tend to spit tungsten into the weld pool when used on AC. Professional welding societies and manufacturers, including the American Welding Society, the Aluminum Association, Miller Electric, and the Lincoln Electric Arc Welding Foundation, do not recommend thoriated tungsten for AC welding. Ceriated or 2% lanthanated tungsten are the types of tungsten recommended for use.

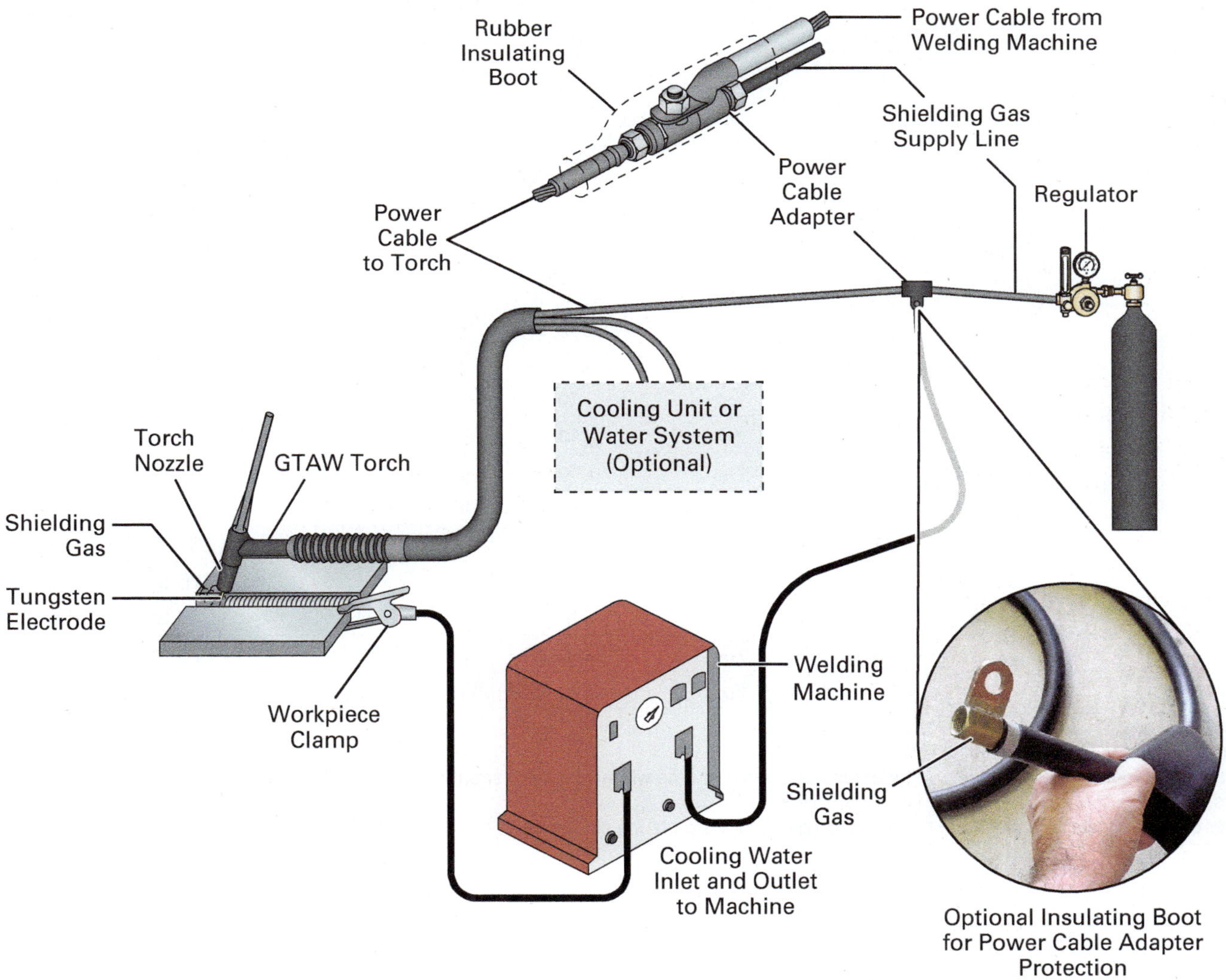

Figure 14 Configuration diagram of typical GTAW welding equipment.

1.4.1 Selecting Shielding Gas

Shielding gas is selected according to the material being welded. Argon is the most commonly used shielding gas. Argon can be used for welding a wide range of materials, including aluminum, titanium, steel, and stainless steel.

Helium and helium-argon mixtures raise the temperature of the arc. This rise in temperature promotes faster welding speeds and deeper weld penetration. However, helium is expensive and can cause difficulty in starting the arc.

Impurities in the shielding gas or mixture with other gases, such as carbon dioxide (CO_2), will contaminate the weld and change the color of the aluminum (usually to black).

CAUTION

Always use the correct gas when performing GTAW. Accidental use of a gas mix intended for GMAW can cause rapid destruction of the tungsten and contamination of the weld.

1.4.2 GTAW Welding Parameters for Aluminum Plate

Different welding techniques and settings are used for different welding situations. The tables in this section provide the common parameters for aluminum welding in each respective position. The specifications for each weld are determined by several factors. *Table 2* shows the parameters for aluminum GTAW fillet welds and *Table 3* shows the parameters for groove welds. Both tables are taken from Hobart Brothers Company. This table only shows a portion of the chart. To access the full table and others like it, access the NCCER Digital Resources page using the QR code at the beginning of this module.

TABLE 2 Parameters for Making GTAW Fillet Welds on Aluminum Plate

Aluminum Thickness (inches)	Weld Position[a]	Preheat[b] (°F)	Weld Passes[c]	Filler Diameter (inches)	Tungsten Electrode Diameter (inches)	Gas Cup Inside Diameter (inches)	Argon Flow (cfh)	AC (amps)	Arc Travel Speed (IPM)
1/16	F, V, H	None	1	3/32	1/16 to 3/32	3/8	16	70 to 100	8 to 10
	O		1	3/32	1/16 to 3/32	3/8	20	65 to 90	8 to 10
3/32	F	None	1	3/32 to 1/8	1/8 to 5/32	3/8	18	110 to 145	8 to 10
	V, H		1	3/32	3/32 to 1/8	3/8	18	90 to 125	8 to 10
	O		1	3/32	3/32 to 1/8	3/8	20	110 to 135	8 to 10
1/8	F	None	1	1/8	1/8 to 5/32	7/16	20	135 to 175	10 to 12
	V, H		1	1/8	3/32 to 1/8	3/8	20	115 to 145	8 to 10
	O		1	1/8	3/32 to 1/8	7/16	25	125 to 155	8 to 10
3/16	F	None	1	5/32	5/32 to 3/16	1/2	25	190 to 245	8 to 10
	V, H		1	5/32	5/32 to 3/16	1/2	25	175 to 210	8 to 10
	O		1	5/32	5/32 to 3/16	1/2	30	185 to 225	8 to 10
1/4	F	None	1	3/16	3/16 to 1/4	1/2	30	240 to 295	8 to 10
	V, H		1	3/16	3/16	1/2	30	220 to 265	8 to 10
	O		1	3/16	3/16	1/2	35	230 to 275	8 to 10
3/8	F	Optional; up to 250°F Max.	2	3/16	1/4	5/8	35	325 to 375	8 to 10
	V		2	3/16	3/16 to 1/4	5/8	35	280 to 315	8 to 10
	H		3	3/16	3/16 to 1/4	5/8	35	270 to 300	8 to 10
	O		3	3/16	3/16 to 1/4	5/8	40	290 to 335	8 to 10

Notes:

[a]F = flat; V = vertical; H = horizontal; O = overhead

[b]Preheating at excessive temperatures or for extended periods of time will reduce weld strength. This is particularly true for alloys in heat-treated tempers.

[c]Number of weld passes and electrode consumption given for weld on one side only.

TABLE 3 Parameters for Making GTAW Groove Welds on Aluminum Plate

Aluminum Thickness (inches)	Weld Position	Edge Prep.	Root Opening (inches)	Preheat (°F)	Weld Passes	Filler Diameter (inches)	Tungsten Electrode Diameter (inches)	Gas Cup Inside Diameter (inches)	Argon (cfh)
$\frac{1}{16}$	F, V, H	A or B	0 to $\frac{1}{16}$	None	1	$\frac{3}{32}$	$\frac{1}{16}$ to $\frac{3}{32}$	$\frac{3}{8}$	20
	O	A or B	0 to $\frac{1}{16}$		1	$\frac{3}{32}$	$\frac{1}{16}$	$\frac{3}{8}$	25
$\frac{3}{32}$	F	A or B	0 to $\frac{3}{32}$	None	1	$\frac{1}{8}$	$\frac{3}{32}$ to $\frac{1}{8}$	$\frac{3}{8}$	20
	V, H	A or B	0 to $\frac{3}{32}$		1	$\frac{3}{32}$ to $\frac{1}{8}$	$\frac{3}{32}$	$\frac{3}{8}$	20
	O	A or B	0 to $\frac{3}{32}$		1	$\frac{3}{32}$ to $\frac{1}{8}$	$\frac{3}{32}$ to $\frac{1}{8}$	$\frac{3}{8}$	25
$\frac{1}{8}$	F	A or B	0 to $\frac{1}{8}$	None	1 to 2	$\frac{1}{8}$ to $\frac{5}{32}$	$\frac{1}{8}$	$\frac{7}{16}$	20
	V, H	A or B	0 to $\frac{3}{32}$		1 to 2	$\frac{1}{8}$	$\frac{1}{8}$	$\frac{7}{16}$	20
	O	A or B	0 to $\frac{3}{32}$		1 to 2	$\frac{1}{8}$ to $\frac{5}{32}$	$\frac{1}{8}$	$\frac{7}{16}$	25

1.5.0 GTAW Terminations, Restarts, and Beads

Before aluminum welding practice begins, it is important to understand how to properly handle restarts and terminations, as well as how to create weave beads and overlapping beads to build a pad.

WARNING!

Unlike steels, aluminum does not change color when heated. Due to its high thermal conductivity, aluminum will get very hot quickly and not be visibly noticeable. This lack of color change poses a workplace danger. To account for the difficulty in distinguishing hot from room temperature aluminum, *OSHA Standard 29 CFR 1910.252(b)(4)(vii)* requires welders to mark the hot metal or provide some other means of warning other workers.

1.5.1 Weld Terminations

A weld termination is made at the end of a weld and usually when the filler metal rod has been consumed. A termination normally results in a crater that must be filled. Most welding codes require the crater to be filled to the full cross section of the weld to prevent crater cracking.

Filling it to this level can be difficult. Most terminations are at the edge of a plate where welding heat tends to build up. This makes it harder to fill the crater. Filling the crater is easier when remote control equipment with a **potentiometer** is used. The potentiometer allows the welding current to be reduced manually as the end of the weld is approached. The control can be foot-operated or located on the torch handle.

Potentiometer: A three-terminal resistor with an adjustable center connection; used for variable voltage control.

In addition to using a manually operated potentiometer to reduce weld current, some welding machines are equipped with a downslope control. This control is set to automatically reduce weld current over a specific time period. The timing of the downslope setting is initiated through a switch on the torch. The welder simply activates the downslope control timing cycle as the termination begins, and the machine begins to reduce the welding current as programmed.

Follow these steps to make a termination using equipment with a potentiometer for current control:

Step 1 As you approach the end of the weld, slowly reduce the welding current. The amount of reduction in welding current is determined by the width of the weld puddle. If the puddle begins to widen, there is too much heat. If the puddle starts to narrow, there is not enough heat.

Step 2 Continue to reduce the welding current while adding filler metal until the crater has been filled.

Step 3 Stop the arc and hold the torch in place, allowing the post-flow of gas to protect both the weld metal and electrode as they cool. The end of the filler metal rod should also remain in the gas flow.

Follow these steps to make a weld termination without remote control equipment:

Step 1 As you approach the end of the weld, add filler metal at a faster rate. The filler metal will absorb excess heat.

Step 2 Continue to add filler metal until the crater has been filled, and then stop the arc.

Step 3 Hold the torch in place, allowing the post-flow of gas to protect both the weld metal and electrode as they cool. The end of the filler metal rod should also remain in the gas flow.

Step 4 Inspect the weld termination for the following characteristics:

- Crater filled to the full cross section of the weld
- No crater cracks
- No excessive porosity with no pores larger than $^3/_{32}$" (2.4 mm)
- No undercut greater than $^1/_{32}$" (0.8 mm) deep or 10% of the base metal thickness, whichever is less

CAUTION

Do not remove the torch or filler metal from the torch shielding gas flow until the puddle has solidified and cooled enough to be unaffected by the atmosphere. The gas post-flow that continues after the welding has stopped protects the molten metal and the electrode. If the torch and its shielding gas flow are removed from the weld area before the weld has solidified, crater porosity or cracks can occur. Also, the end of the filler metal will be contaminated, and the contaminated end will have to be discarded. Finally, the tungsten electrode can oxidize (turn a bluish-black color), forcing redressing of the electrode.

1.5.2 Weld Restarts

A restart is the junction where a new weld connects to the bead of a previous weld to continue the weld. A restart must be made so that it blends smoothly with the previous weld and does not stand out. An improperly made restart will create a weld discontinuity. Whenever possible, avoid restarts by running a bead the full length of the weld joint.

A properly made restart blends into the bead and is hard to detect. If the restart has undercut, not enough time was spent in the crater to fill it. If the undercut is on one side or the other, use more of a side-to-side motion as you move back into the crater. If the restart has a lump, it was overfilled; too much time was spent in the crater before the forward motion was resumed.

Follow these steps to make a restart:

Step 1 Clean the area of the restart using an approved cleaning method.

Step 2 Hold the torch at the proper angle and arc distance. Restart the arc directly over the center of the crater. Remember that welding codes do not allow arc strikes outside of the area that is to be welded.

Step 3 Move the electrode tip in a small circular motion over the crater and add filler metal when the molten puddle is the same size as the crater.

Step 4 As soon as the puddle fills the crater, advance the puddle slightly. Continue to add filler metal as needed.

Step 5 Inspect the restart.

Step 6 Continue to practice restarts until they are correct.

Thin Materials

When welding thin materials, stop about 1" or 2" (25.7 mm or 50 mm) from the edge without concern for the crater. Then back weld from the edge back to the crater and fill it as described.

1.5.3 Weave Beads

The two basic bead types used in GTAW are weave beads and stringer beads. Stringer and weave beads are used to make welds, while overlapping stringer and weave beads are used to build pads. Always check the WPS or site quality standards to determine whether stringer or weave beads should be used.

Weave beads (*Figure 15*) are made by working the weld puddle back and forth across the axis of the weld. This produces a weld bead much wider than the puddle width. The width of a weave bead is determined by the amount of oscillation or cross motion. Weave beads put more heat into the base metal.

When making a weave bead, take special care at the toes to ensure proper fusion to the base metal. Do this by slowing down or pausing slightly at the edges. Pausing at the edges will also flatten out the weld and give it the proper profile. When making weave beads on strengthened aluminums, including heat-treated and strain-hardened, these weave beads lose strength when welded. Some aluminum alloys lose strength more easily than others. Due to these risks, weave beads on aluminum should be avoided to minimize the loss of strength caused by the heat input from the welding arc. In practice, only annealed aluminum can be welded without experiencing weakening.

Practice weave beads in the flat position first. Experiment with different weave motions and push angles.

Follow these steps to create weave beads:

Step 1 Hold the torch at the desired angle with the electrode tip directly over the point where the weld will begin.

Step 2 Lower the torch and filler rod to the weld.

Step 3 Hold the arc in place until the weld puddle begins to form. Add filler metal.

Step 4 Slowly advance the arc in a weaving motion while maintaining the torch angle and adding filler metal.

Step 5 Continue to weld until a bead about 2" to 3" (5 cm to 8 cm) long is formed.

Step 6 Stop and then fill the crater while slowly reducing amperage with a current control to allow the weld puddle to solidify. Turn off the current. Hold the filler rod and torch in place until both the tungsten electrode and weld cool under the protection of the gas envelope.

Step 7 Inspect the bead for the following characteristics:

- Straightness of the bead
- Uniform appearance of the bead face; face and root reinforcement depth in accordance with applicable codes
- Smooth, flat transition with complete fusion at the toes of the weld
- Termination crater filled
- No excessive porosity with no pores larger than $^3/_{32}$" (2.4 mm)
- No undercut greater than $^1/_{32}$" (0.8 mm) deep or 10% of the base metal thickness, whichever is less
- No overlap
- No overfill
- No cracks

Step 8 Continue practicing weave beads until you can make acceptable beads every time.

CAUTION

Do not exceed the recommended weave bead width. The width of weave beads is often specified in the welding code or WPS used at the site.

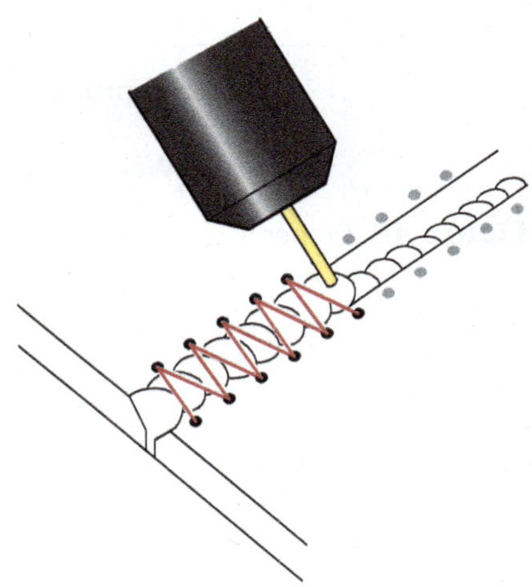

(A)

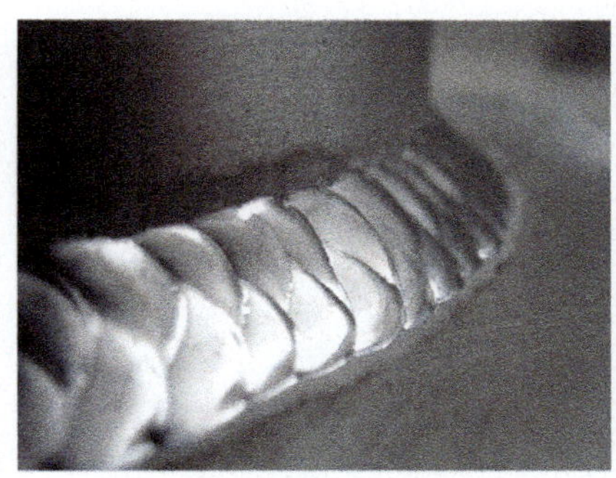

(B)

Figure 15 Weave bead.
Source: Photo courtesy of The Lincoln Electric Company, Cleveland, OH, U.S.A. (15B)

Cooling Practice Coupons

Aluminum practice coupons can be cooled with water. Remember that the maximum interpass temperature of aluminum is typically 250°F (121°C). If water is used to quench coupons, it is essential that the workpiece be thoroughly dried before welding resumes. Any moisture on the surface causes hydrogen contamination of the weld.

WARNING!

Use pliers and gloves to handle hot practice coupons. Gloves should remain on when placing hot practice coupons in water. Steam will rise off the coupons and can burn or scald unprotected hands.

1.5.4 Overlapping Beads

Overlapping beads are made by depositing connective weld beads parallel to one another. Each successive bead overlaps the toe of the previous bead to form a raised, flat surface. This is also called *padding*. Overlapping beads are used to build up surfaces and to make multipass welds. Both stringer and weave beads can be overlapped. Properly overlapped beads, when viewed from their ends, form a nearly flat, level surface (*Figure 16*).

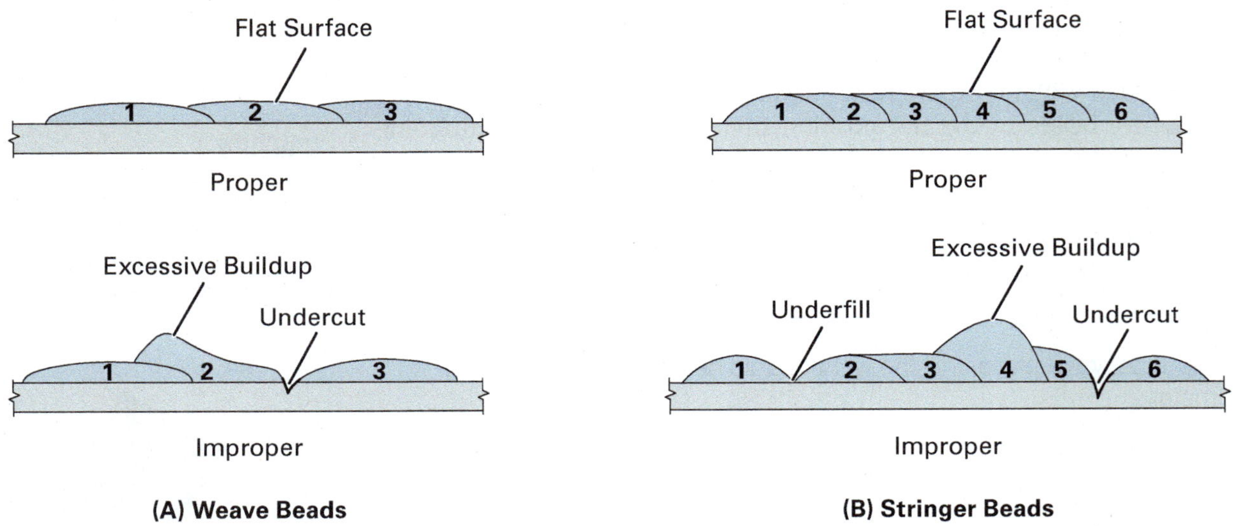

Figure 16 Proper and improper overlapping beads.

Follow these steps to make GTAW overlapping stringer beads using $^1/_8$" (3.2 mm) aluminum filler metal:

Step 1 Mark out a square on a piece of aluminum.

Step 2 Deoxidize the base metal.

Step 3 Weld a stringer bead along one edge.

Step 4 Clean the bead with a stainless steel brush. Make sure that the base metal is cleaned with the brush between each pass.

Step 5 Run a second bead along the previous bead. Be sure to overlap the previous bead to obtain a good fusion and produce a flat profile.

Step 6 Continue running overlapping stringer beads until the face of the square piece of metal has been covered.

Step 7 Continue building layers of stringer beads (pads), one on top of the other, until your technique has been perfected.

Continue to weld overlapping beads. Build a pad using stringer and weave beads as shown in *Figure 17*.

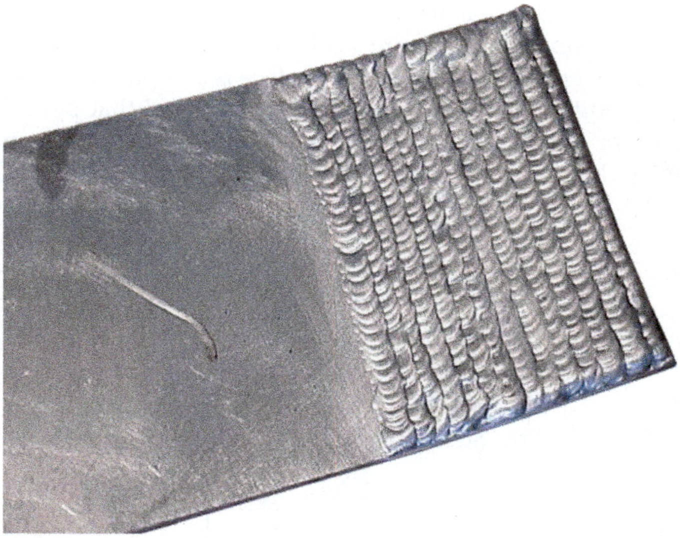

Figure 17 Pad built with stringer and weave beads.
Source: Terry Lowe

1.6.0 Fillet Welds

Fillet welds generally require less base metal preparation since bevels are not needed. However, the cleaning process must be thorough.

The most common fillet welds are made in lap and T-joints. The weld position for plate is determined by the axis of the weld and the orientation of the workpiece. The positions for fillet welding, shown in *Figure 18*, are flat (1F), horizontal (2F), vertical (3F), and overhead (4F). In the 1F and 2F positions, the weld axis can be inclined up to 15°. Any weld axis inclination for the other positions varies with the rotational position of the weld face as specified in AWS standards.

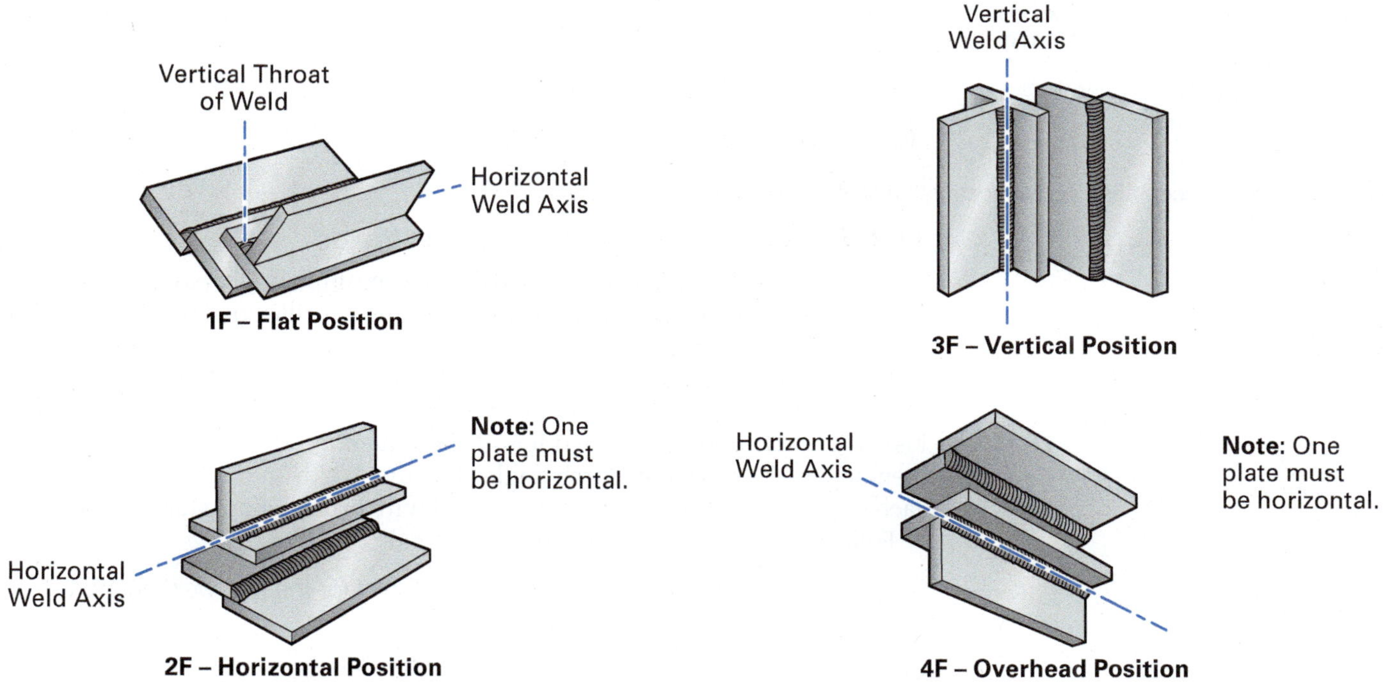

Figure 18 Fillet welding positions.

T-Joint Heat Dissipation

In T-joints, the welding heat dissipates more rapidly in the thicker or non-butting member. On the root pass, the arc may have to be slightly more concentrated on the thicker or the non-butting member to compensate for heat loss.

Fillet welds can be concave or convex, depending on the WPS or site quality standards. The welding codes require a fillet weld to have a uniform concave or convex face. However, a slightly nonuniform face is acceptable. The convexity of a fillet weld or individual surface bead must not exceed that permitted by the applicable code or standard. A fillet weld with profile defects is unacceptable and must be repaired. Ideal, acceptable, and unacceptable profiles are shown in *Figure 19*.

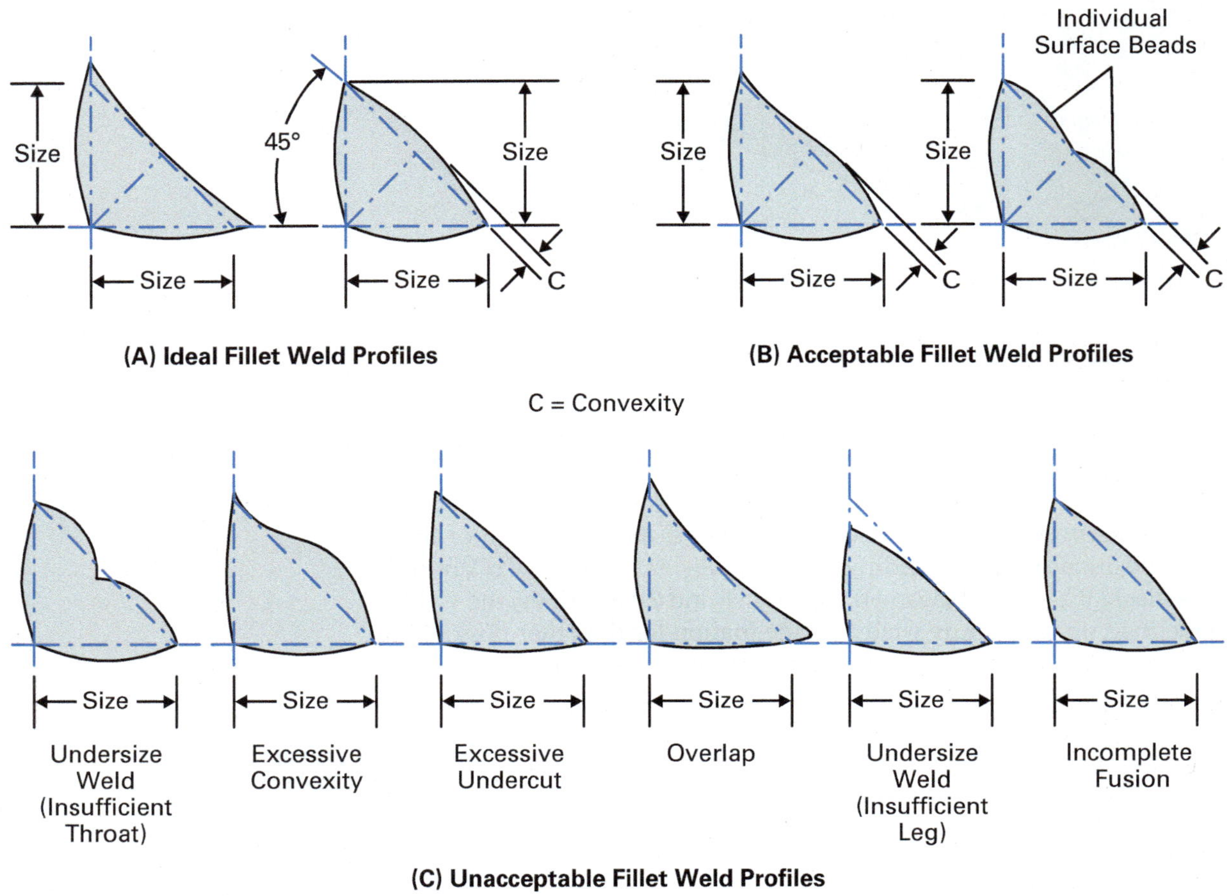

Figure 19 Ideal, acceptable, and unacceptable fillet weld profiles.

1.6.1 Practicing Flat (1F) Position Fillet Welds

Practice flat (1F) position fillet welds. Make multipass (six-pass) convex fillet welds in a T-joint using an appropriate aluminum filler metal. When making flat fillet welds, pay close attention to the torch work angle and travel speed. For the first bead, the work angle should be approximately 45° to both plate surfaces, slightly favoring the non-butting member to accommodate its greater heat loss. The work angle is adjusted for all subsequent beads for proper bead placement. *Figure 20* shows the torch angles for fillet welds in the 1F position. Clean each completed bead with a stainless steel wire brush before starting the next.

Follow these steps to make a flat fillet weld:

Step 1 Tack two plates together to form a T-joint for the fillet weld coupon (*Figure 21*). Clean the tack welds.

Step 2 Clamp or tack-weld the coupon in the 1F position (flat).

Step 3 Run the first bead along the root of the joint. Use a work angle of approximately 45° and a 15° to 20° push angle.

Step 4 Run the second bead along a toe of the first weld, overlapping about 75% of the first bead. Alter the work angle as shown in *Figure 20*. Use a 15° to 20° push angle with a slight oscillation.

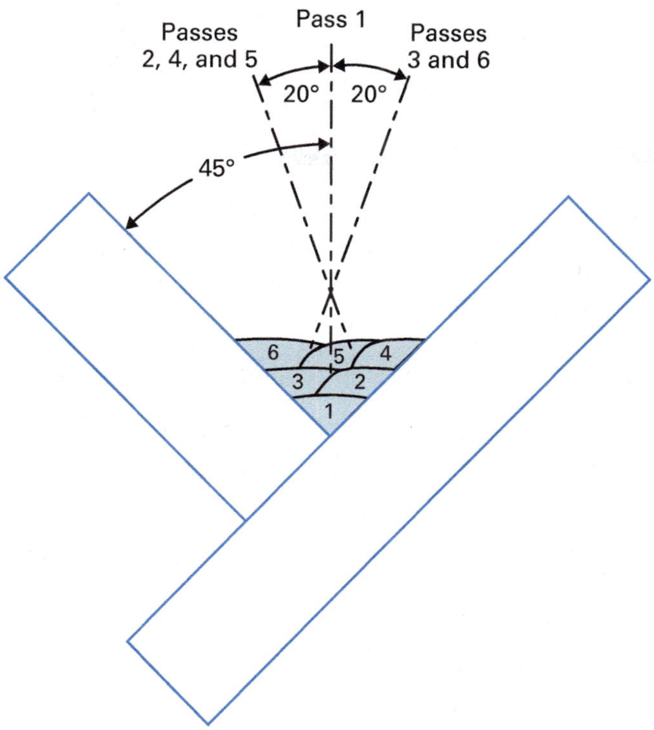

Figure 20 Multipass 1F bead sequence and work angles.

Step 5 Run the third bead along the other toe of the first weld. Be sure to fill the groove created when the second bead was run. Alter the work angle as shown in *Figure 20* and use a 15° to 20° push angle with a slight oscillation.

Step 6 Run the fourth bead along the outside toe of the second weld, overlapping about half of the second bead. Alter the work angle for proper bead placement and use a 15° to 20° push angle with a slight oscillation.

Step 7 Run the fifth bead along the inside toe of the fourth weld, overlapping about half of the fourth bead. Alter the work angle for proper bead placement and use a 15° to 20° push angle with slight oscillation.

Step 8 Run the sixth bead along the toe of the fifth weld. Be sure to fill the groove created when the fifth bead was run. Alter the work angle for proper bead placement and use a 15° to 20° push angle with slight oscillation.

Step 9 Have your instructor inspect the weld. The weld is acceptable if it exhibits the following characteristics:

- Uniform appearance of the bead face
- Craters and restarts filled to the full cross section of the weld
- Uniform weld size of $\pm {}^{1}/_{16}$" (±1.6 mm)
- Acceptable weld profile in accordance with *Figure 19* of this module
- Smooth transition with complete fusion at the toes of the weld
- No excessive porosity with no pores larger than ${}^{3}/_{32}$" (2.4 mm)
- No undercut greater than ${}^{1}/_{32}$" (0.8 mm) deep or 10% of the base metal thickness, whichever is less
- No overlap
- No inclusions
- No cracks
- No concavity

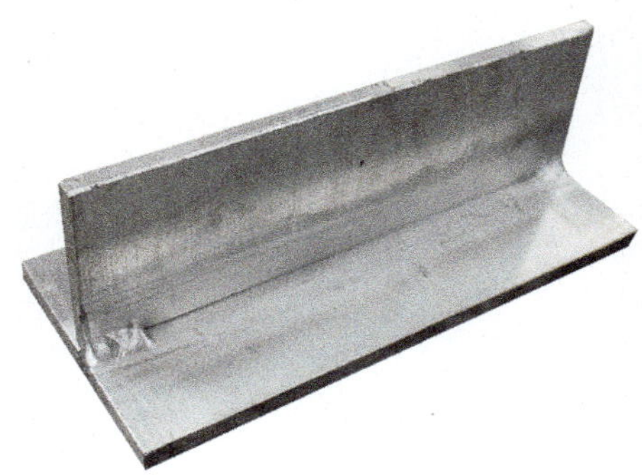

Figure 21 Fillet weld coupon.
Source: Terry Lowe

Raising the Working Height of a GTAW Torch

In some cases, slightly raising the torch height (increasing the arc length) can be used to widen the weld beads, instead of relying on the technique of oscillating the torch.

1.6.2 Practicing Horizontal (2F) Position Fillet Welds

Practice horizontal (2F) fillet welding by placing multipass fillet welds in a T-joint. Use an appropriate filler metal as directed by your instructor. When making horizontal fillet welds, pay close attention to the torch angles. For the first bead, the work angle should be approximately 45° to both plate surfaces, slightly favoring the non-butting member to accommodate its greater heat loss. The work angle is adjusted for all other passes.

Follow these steps to make a horizontal fillet weld:

Step 1 Tack two plates together to form a T-joint for the fillet weld coupon. Clean the tack welds.

Step 2 Clamp or tack-weld the coupon in the horizontal position.

Step 3 Run the first bead along the root of the joint. Use a work angle of about 45° and a 15° to 20° push angle (*Figure 22*).

Step 4 Clean the weld.

Remaining passes: A nonstandard term used to describe the fill and cover passes as a group.

Step 5 Run the **remaining passes** at the proper work angles using a 15° to 20° push angle and slight oscillation. Overlap each previous pass. Clean the weld after each pass.

Step 6 Have your instructor inspect the weld. The weld is acceptable if it exhibits the following characteristics:

- Uniform appearance on the bead face
- Craters and restarts filled to the full cross section of the weld
- Uniform weld size of $\pm^{1}/_{16}$" (±1.6 mm)
- Acceptable weld profile in accordance with *Figure 19* of this module
- Smooth transition with complete fusion at the toes of the weld
- No excessive porosity with no pores larger than $^{3}/_{32}$" (2.4 mm)
- No undercut greater than $^{1}/_{32}$" (0.8 mm) deep or 10% of the base metal thickness, whichever is less
- No overlap
- No inclusions
- No cracks
- No concavity

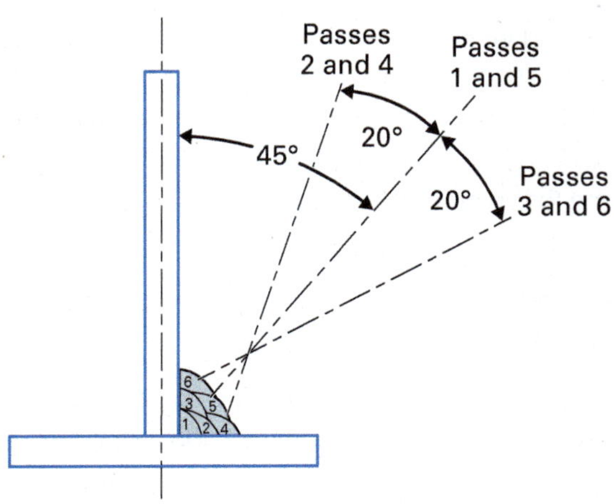

Figure 22 Multipass 2F bead sequence and work angles.

1.6.3 Practicing Vertical (3F) Position Fillet Welds

Practice vertical (3F) fillet welding by placing multipass fillet welds in a T-joint. Vertical welds are usually made by welding uphill from the bottom to the top using a push angle. For the first bead, the work angle should be approximately 45° to both plate surfaces, slightly favoring the non-butting member to accommodate its greater heat loss. Because of the uphill welding and push angle, this type of weld is sometimes called *vertical-up fillet welding*.

Either stringer or weave beads can be used for vertical welding. On the job, the site WPS or site quality standard will specify which technique to use.

Follow these steps to make an uphill fillet weld:

Step 1 Tack two plates together to form a T-joint for the fillet weld coupon.

Step 2 Clamp or tack-weld the coupon in the vertical position.

Step 3 Run the first bead along the root of the joint, starting at the bottom. Use a work angle of about 45° and a 15° to 20° push angle (*Figure 23*). Pause in the weld puddle to fill the crater.

Step 4 Clean the weld.

Step 5 Run the remaining passes using a 45° work angle and a 15° to 20° push angle with a side-to-side weave technique. Use a slow motion across the face of the weld. Pause at each toe for penetration and to fill the crater. Clean the weld after each pass.

Step 6 Have your instructor inspect the weld. The weld is acceptable if it exhibits the following characteristics:

- Uniform appearance on the bead face
- Craters and restarts filled to the full cross section of the weld
- Uniform weld size of ±$\frac{1}{16}$" (±1.6 mm)
- Acceptable weld profile in accordance with *Figure 19* of this module
- Smooth transition with complete fusion at the toes of the weld
- No excessive porosity with no pores larger than $\frac{3}{32}$" (2.4 mm)
- No undercut greater than $\frac{1}{32}$" (0.8 mm) deep or 10% of the base metal thickness, whichever is less
- No overlap
- No inclusions
- No cracks
- No concavity

Repeat vertical fillet (3F) welding using stringer beads and a slight side-to-side motion. Pause briefly at each toe to prevent undercut. For stringer beads, use a 15° to 20° push angle and adjust the work angle to properly position each bead.

NOTE

Check with your instructor to see if you should run stringer beads or weave beads, or practice both techniques.

1.6.4 Practicing Overhead (4F) Position Fillet Welds

Practice overhead (4F) position fillet welds by making multipass fillet welds in a T-joint. When making overhead fillet welds, pay close attention to the work angle. For the first bead, the work angle should be approximately 45° to both plate surfaces, slightly favoring the non-butting member to accommodate its greater heat loss. The work angle is adjusted for all other passes.

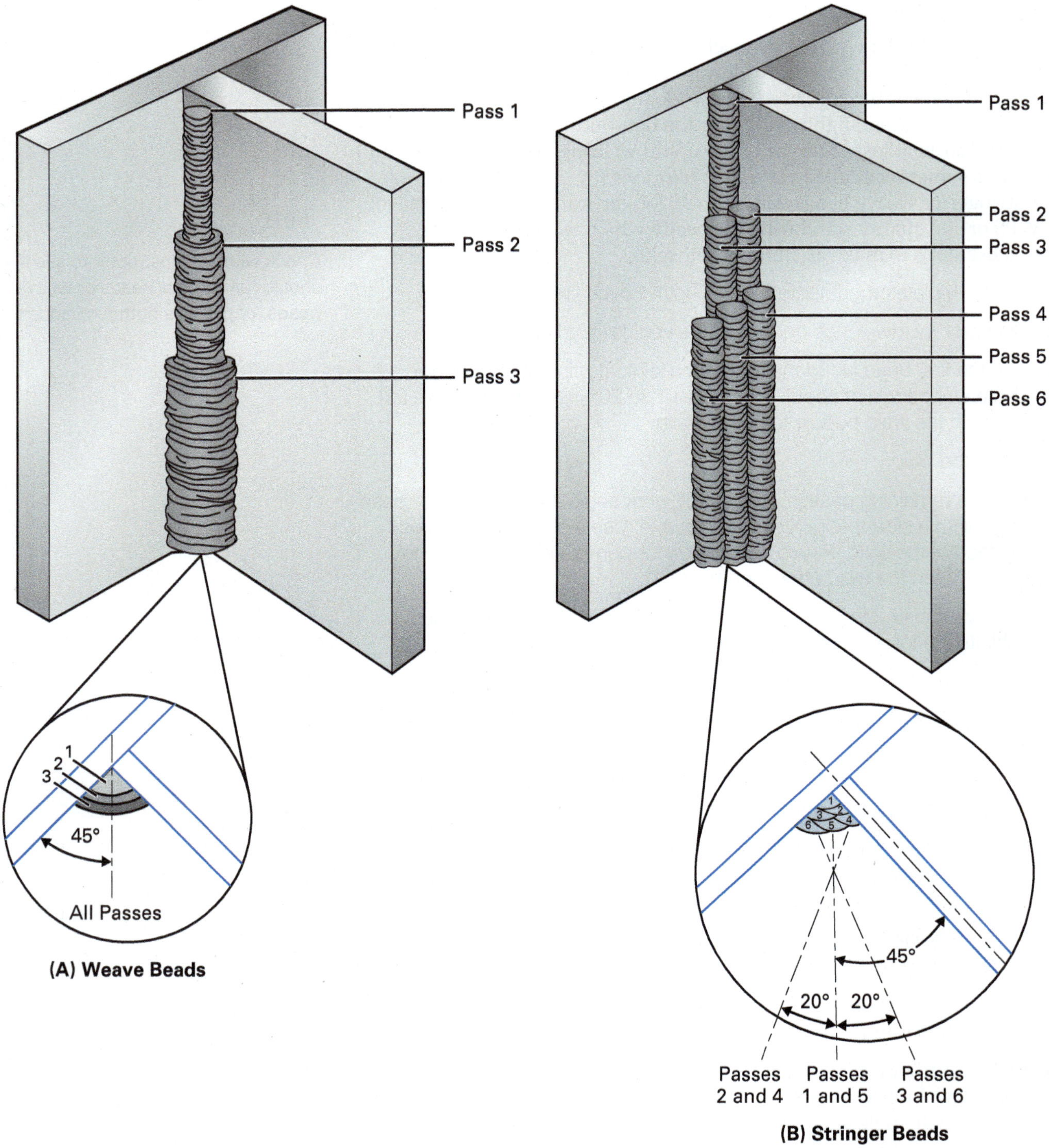

Figure 23 Multipass 3F bead sequences and work angles for stringer and weave beads.

Follow these steps to make an overhead fillet weld:

Step 1 Tack two plates together to form a T-joint for the fillet weld coupon.

Step 2 Clamp or tack-weld the coupon so that it is in the overhead position.

Step 3 Run the first bead along the root of the joint. Use a work angle of about 45° and a 15° to 20° push angle.

Step 4 Clean the weld.

Step 5 Run the remaining passes at the proper work angles using a 15° to 20° push angle and slight oscillation (*Figure 24*). Overlap each previous pass. Clean the weld after each pass.

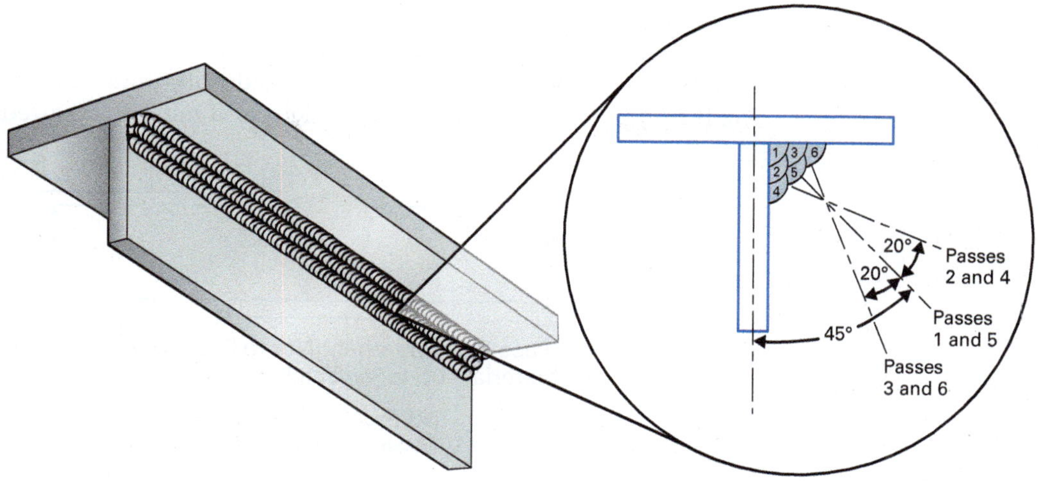

Figure 24 Multipass 4F bead sequence and work angles.

Step 6 Have your instructor inspect the weld. The weld is acceptable if it exhibits the following characteristics:

- Uniform appearance on the bead face
- Craters and restarts filled to the full cross section of the weld
- Uniform weld size of ±1⁄16" (±1.6 mm)
- Acceptable weld profile in accordance with *Figure 19* of this module
- Smooth transition with complete fusion at the toes of the weld
- No excessive porosity with no pores larger than 3⁄32" (2.4 mm)
- No undercut greater than 1⁄32" (0.8 mm) deep or 10% of the base metal thickness, whichever is less
- No overlap
- No inclusions
- No cracks
- No concavity

1.7.0 V-Groove Plate Welds

The V-groove weld is a common groove weld used on both plate and pipe. The backing method of welding an aluminum V-groove joint with GTAW is presented in this module. Practicing V-groove welds on plate is good preparation for making the more difficult pipe welds.

For the **root pass**, use a 1⁄8" (3.2 mm) aluminum filler metal rod. After the root pass has been completed, clean and inspect it. Use the appropriate cleaning method to remove the oxidation that tends to form along the joint. Inspect the root pass for the following criteria:

Root pass: The initial weld pass made to produce a root bead.

- Uniform, smooth face
- No excessive buildup
- No undercut greater than 1⁄32" (0.8 mm) deep or 10% of the base metal thickness, whichever is less
- No overlap
- No excessive porosity with no pores larger than 3⁄32" (2.4 mm)
- Proper fusion

V-groove welds with backing should be made with slight face reinforcement, not exceeding $^{3}/_{32}$" (2.4 mm) for materials up to $^{3}/_{8}$" (9.6 mm) thick. There should be a gradual transition to the base metal at each toe. Groove welds must not have excessive face reinforcement, any underfill, undercut, or overlap (*Figure 25*). If a groove weld has any of these defects, it must be repaired.

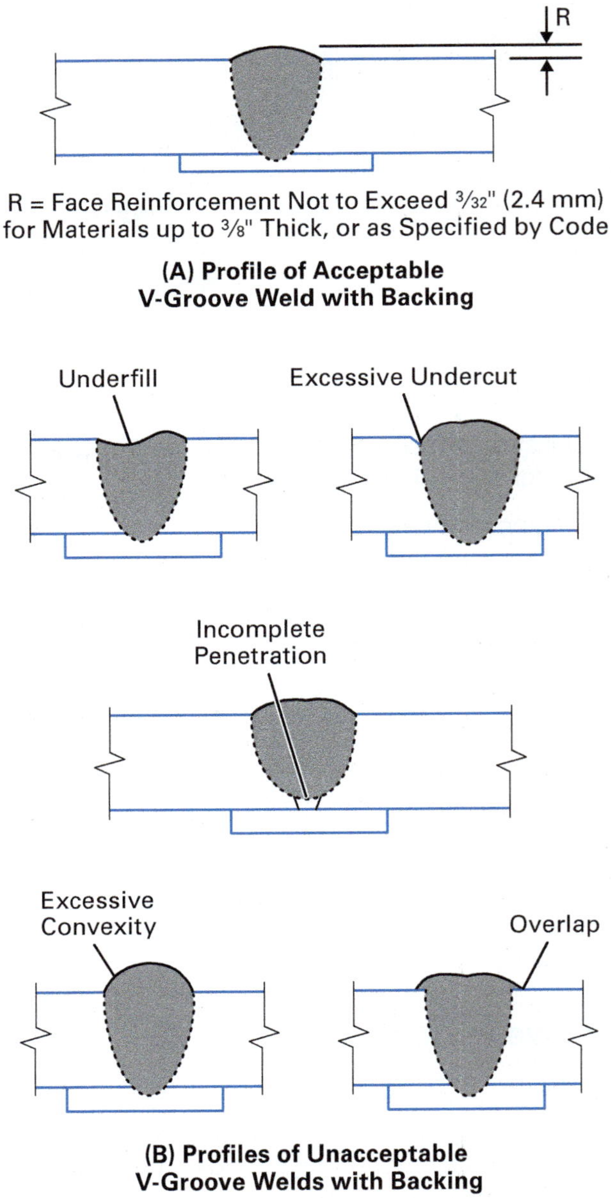

Figure 25 Acceptable and unacceptable V-groove weld profiles.

Modified U-Groove Joint without Backing

Instead of a backing strip, a modified U-groove joint without backing can be used when welding aluminum plate. The modified U-groove joint without backing shown is the same as the one that is used for GTAW of aluminum pipe. Use a degreaser on the back of the joint, but do not brush the metal. Leaving a bit of oxide in place on the back helps support the weld.

Source: Terry Lowe

V-groove welds with backing can be made in all positions (*Figure 26*). The weld position for plate is determined by the axis of the weld and the orientation of the workpiece. Groove weld positions are flat (1G), horizontal (2G), vertical (3G), and overhead (4G).

During the practice sessions that follow, use $\frac{1}{8}$" (3.2 mm) aluminum filler metal for the root as well as the other passes unless your instructor directs otherwise. Pay special attention to filling the crater at the termination of the weld. Clean each completed bead with a stainless steel wire brush before starting the next bead.

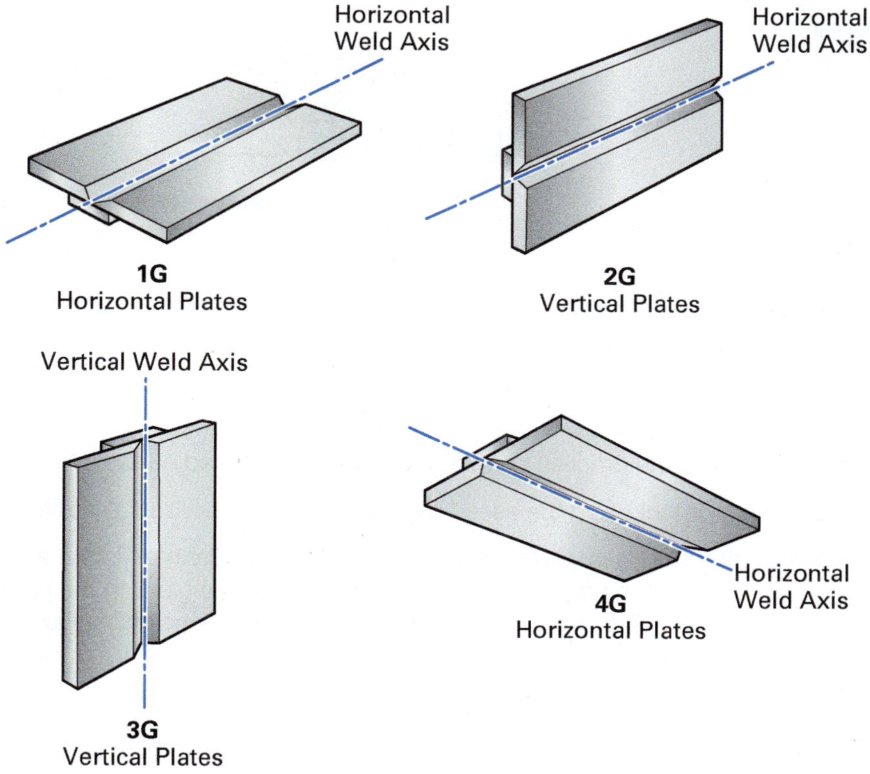

Figure 26 V-groove weld positions.

1.7.1 Practicing Flat (1G) Position V-Groove Welds with Backing

Follow these steps to make V-groove welds with backing in the flat (1G) position (*Figure 27*):

Step 1 Tack-weld the practice coupon together as explained earlier.

Step 2 Clamp or tack-weld the assembled coupon in the flat position above the welding table surface.

Step 3 Use a 15° to 20° push angle and a 0° work angle to weld the root pass with an appropriate aluminum filler metal. Feed the filler metal at a 15° to 20° angle to the plate.

Step 4 Clean the root pass.

Step 5 Run the remaining passes using a 10° to 15° push angle and the bead sequence and work angles shown in *Figure 27*. Hold the filler metal about 15° to 20° above the plate face. Clean the weld between each pass.

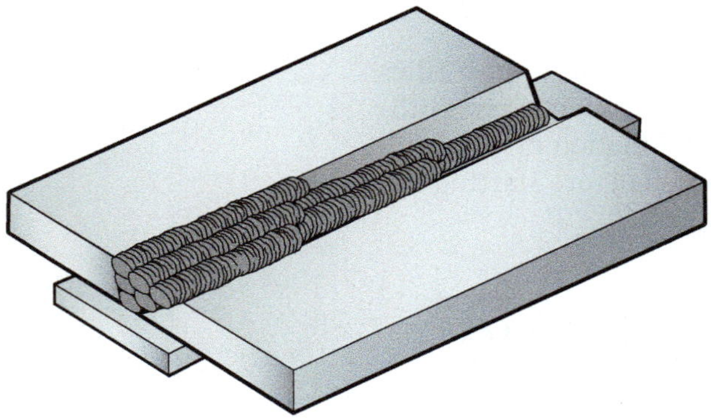

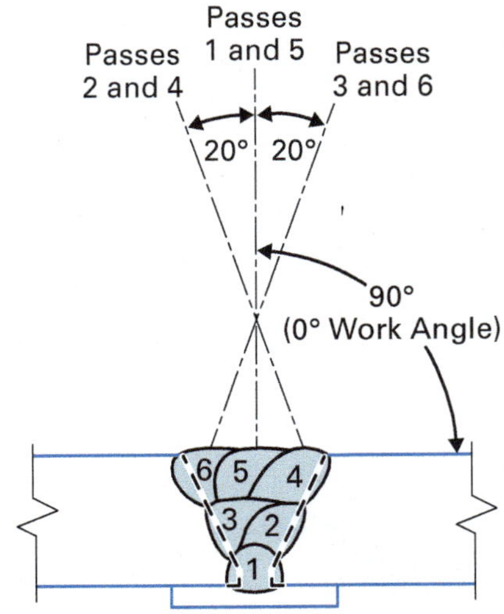

Figure 27 Multipass 1G bead sequence and work angles.

1.7.2 Practicing Horizontal (2G) Position V-Groove Welds with Backing

Follow these steps to practice V-groove welds with backing in the horizontal (2G) position (*Figure 28*):

Step 1 Tack-weld the practice coupon together as described earlier.

Step 2 Clamp or tack-weld the assembled coupon in the horizontal position.

Step 3 Run the root pass with an appropriate aluminum filler metal. Use a 15° to 20° push angle and a 15° to 20° filler metal angle with the plate surface. Use the appropriate work angle for Pass 1 shown in *Figure 28*.

Step 4 Clean the weld.

Step 5 Run the remaining passes using a 15° to 20° push angle and the bead sequence and work angles shown in *Figure 28*. Clean the weld between each pass.

1.7.3 Practicing Vertical (3G) Position V-Groove Welds with Backing

Follow these steps to practice V-groove welds with backing in the vertical (3G) position (*Figure 29*):

Step 1 Tack-weld the practice coupon together as described earlier.

Step 2 Clamp or tack-weld the assembled coupon in the vertical position.

Step 3 Run the root pass uphill using an appropriate aluminum filler metal. Use a push angle of 15° to 20° and a filler metal angle of 15° to 20 ° to the plate surface. Use a 0° work angle for Pass 1 as shown in *Figure 29*.

Step 4 Clean the weld.

Step 5 Run the remaining passes uphill to complete the weld. Use a 15° to 20° push angle and the bead sequence and work angles shown in *Figure 29*. Clean the weld between each pass.

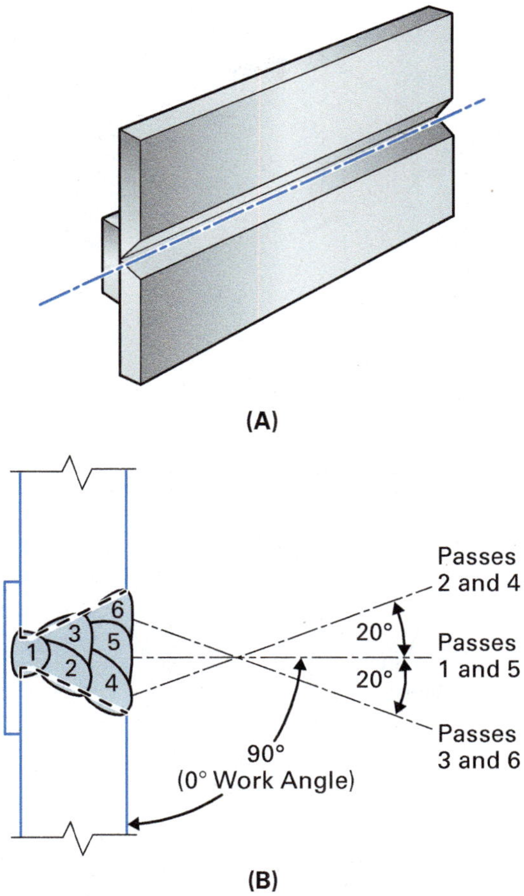

Figure 28 Multipass 2G bead sequences and work angles.

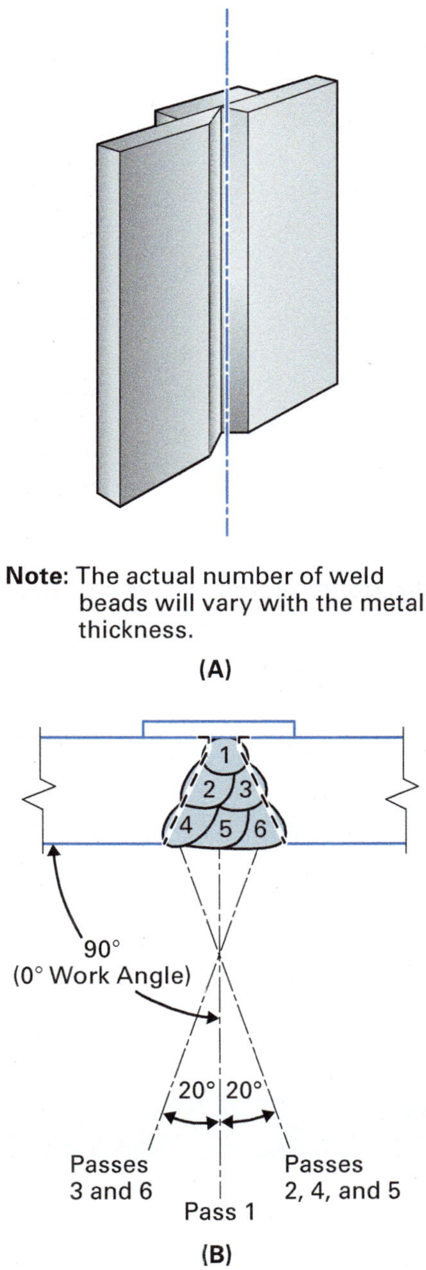

Figure 29 Multipass 3G bead sequence and work angles.

1.7.4 Practicing Overhead (4G) Position V-Groove Welds with Backing

Follow these steps to practice V-groove welds with backing in the overhead (4G) position (*Figure 30*):

Step 1 Tack-weld the practice coupon together as explained earlier.

Step 2 Clamp or tack-weld the coupon in the overhead position.

Step 3 Run the root pass using an appropriate aluminum filler metal. Use a 15° to 20° push angle and a 15° to 20° filler metal angle with the plate surface. Use a 0° work angle for Pass 1, as shown in *Figure 30*.

Step 4 Clean the weld.

Step 5 Run the remaining passes using a 15° to 20° torch push angle and the bead sequence and work angles shown in *Figure 30*. Clean the weld between each pass.

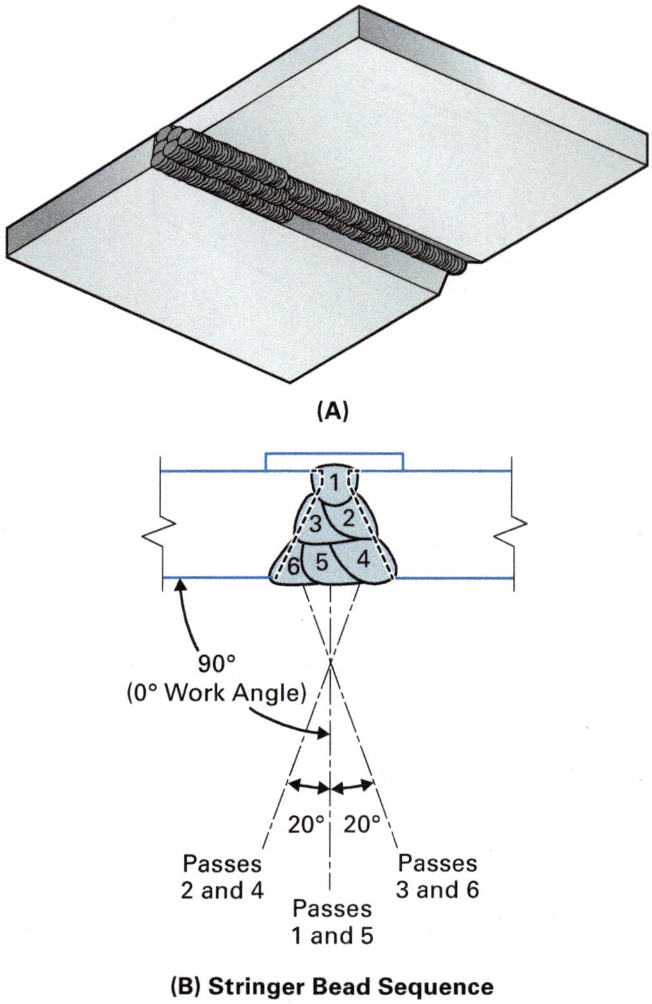

Figure 30 Multipass 4G bead sequence and work angles.

1.0.0 Section Review

1. A slow travel speed in GTAW forms weld puddles that are _____.
   a. small and shallow
   b. large and penetrate deeply
   c. contaminated by the argon
   d. oxidized and gel-like

2. For occasional, short-term welding of materials containing manganese, a welder should wear a _____.
   a. damp bandanna over the mouth and nose
   b. tight-fitting disposable dust mask
   c. full-face, supplied-air respirator
   d. HEPA-rated filter on a respirator

3. The two basic factors to be considered when selecting aluminum filler metals are the base metal composition and the _____.
   a. Welding Procedure Specification (WPS)
   b. equipment manufacturer
   c. base metal manufacturer
   d. Aluminum Association

4. The typical range of frequencies used for aluminum GTAW with an inverter welding machine is _____.
   a. 50 cycles to 60 cycles
   b. 90 cycles to 100 cycles
   c. 120 cycles to 140 cycles
   d. 280 cycles to 300 cycles

5. Filling the crater of an aluminum weld termination is easier when the GTAW torch is equipped with a(n) _____.
   a. potentiometer
   b. phase inverter
   c. wire feeder
   d. oversized tungsten

6. The appropriate travel angle for an aluminum fillet weld in the 4F position is a _____.
   a. 40° to 50° drag angle
   b. 15° to 20° push angle
   c. 40° to 50° push angle
   d. 15° to 20° drag angle

7. Aluminum V-groove welds on $\frac{1}{4}$" plate should have slight face reinforcement, *not* exceeding _____.
   a. $\frac{1}{32}$" (0.8 mm)
   b. $\frac{1}{16}$" (1.6 mm)
   c. $\frac{3}{32}$" (2.4 mm)
   d. $\frac{1}{8}$" (3.2 mm)

Module 29402 Review Questions

1. To maintain the desired puddle size and advance the puddle when using the freehand technique, the welder may _____.
   a. leave the end of the filler rod in the puddle
   b. place the electrode in contact with the puddle
   c. preheat the filler metal rod
   d. move the torch tip in a small circular motion

2. The aluminum filler metal typically used for the GTAW process is a(n) _____.
   a. dual-shielded continuous wire
   b. uncoated, solid metal rod
   c. carbon-coated rod
   d. composite core rod

3. To clean contamination such as oil, dirt, and grease from aluminum filler metal rods, you should use a clean rag and _____.
   a. a citrus-based cleaner
   b. silicone spray
   c. hydrochloric acid
   d. acetone

4. When preparing beveled aluminum plate to be welded with a backing strip, how wide must the root opening be?
   a. $\frac{1}{32}$" (0.8 mm)
   b. $\frac{1}{16}$" (1.6 mm)
   c. $\frac{1}{8}$" (3.2 mm)
   d. $\frac{3}{8}$" (9.6 mm)

5. Helium and helium-argon shielding gas mixtures enable faster welding speeds and deeper penetration because helium _____.
   a. raises the temperature of the arc
   b. lowers the melting point of aluminum
   c. protects a larger area than argon alone
   d. combines with oxygen to eliminate oxidation

6. Filling the crater of a weld termination is made easier by _____.
   a. slowly increasing the welding current
   b. slowly decreasing the welding current
   c. rapidly increasing the arc length
   d. rapidly decreasing the arc length

7. When welding weave beads, to ensure proper fusion to the base metal at the toes of the weld, the welder should _____.
   a. decrease the arc length
   b. change to a drag angle
   c. stay well away from the toes
   d. slow down or pause briefly

8. The weld position in a fillet weld is determined by the orientation of the workpiece and the _____.
   a. size of the electrode
   b. axis of the weld
   c. type of joint
   d. welder's preference

9. In the 2F position, the *first* bead of the joint should be made with a work angle of approximately _____.
   a. 20°
   b. 45°
   c. 60°
   d. 70°

10. Aluminum plate V-groove welds in the 2G position should be made using a filler metal rod angle of _____.
   a. 90°
   b. 45°
   c. 15° to 20°
   d. 5° to 10°

Answers to odd-numbered questions are found in the Review Question Answer Keys at the back of this book.

Answers to Section Review Questions

Answer	Section Reference	Objective
Section 1.0.0		
1. b	1.2.1	1b
2. d	1.1.4	1a
3. a	1.3.1	1c
4. c	1.4.0	1d
5. a	1.5.1	1e
6. b	1.6.4	1f
7. c	1.7.0	1g

MODULE 29403

GTAW – Aluminum Pipe

Source: Photo courtesy of The Lincoln Electric Company, Cleveland, OH, U.S.A.

Objective

Successful completion of this module prepares you to do the following:

1. Explain how to prepare for GTAW aluminum pipe welding and describe specific pipe welding techniques.
   a. Identify common welding safety practices.
   b. Describe common GTAW pipe welding techniques.
   c. Explain how to prepare the area and materials for aluminum GTAW practice.
   d. Explain how to prepare GTAW equipment for welding.
   e. Describe techniques and positions related to aluminum pipe GTAW.
   f. Describe how to make U-groove GTAW welds on aluminum pipe.

Performance Task

Under supervision, you should be able to do the following:

1. Make GTAW U-groove welds on aluminum pipe in the 2G, 5G, and 6G positions.

Overview

The gas tungsten arc welding (GTAW) process is unique in that the welder must add the filler metal using the opposite hand. The process is well suited for welding aluminum, using the freehand technique. Welders must understand that a significant amount of welding current is required for welding aluminum, especially when working with thick materials. This module provides insight into welding aluminum pipe using the GTAW process and the U-groove joint, which does not require a metal backing or a backing gas.

NCCER Industry-Recognized Credentials

If you are training through an NCCER-accredited sponsor, you may be eligible for credentials from NCCER. The ID number for this module is 29403. Note that this module may have been used in other NCCER curricula and may apply to other level completions. Contact NCCER at 1.888.622.3720 or go to **www.nccer.org** for more information.

You can also show off your industry-recognized credentials online with NCCER's digital credentials. Transform your knowledge, skills, and achievements into credentials that you can share across social media platforms, send to your network, and add to your resume. For more information, visit **www.nccer.org**.

Digital Resources for Welding

SCAN ME

Scan this code using the camera on your phone or mobile device to view the digital resources related to this craft.

1.0.0 Aluminum Pipe GTAW

Performance Task

1. Make GTAW U-groove welds on aluminum pipe in the 2G, 5G, and 6G positions.

Objective

Explain how to prepare for GTAW aluminum pipe welding and describe specific pipe welding techniques.

a. Identify common welding safety practices.
b. Describe common GTAW pipe welding techniques.
c. Explain how to prepare the area and materials for aluminum GTAW practice.
d. Explain how to prepare GTAW equipment for welding.
e. Describe techniques and positions related to aluminum pipe GTAW.
f. Describe how to make U-groove GTAW welds on aluminum pipe.

Gas tungsten arc welding (GTAW) uses an arc between a tungsten electrode and the base metal to melt the base and filler metals. The electrode, the arc, and the molten base metal are shielded from atmospheric contamination by a flow of inert gas from the torch nozzle (*Figure 1*). The filler metal is a handheld rod compatible with the base metal. The filler metal is manually fed into the leading edge of the weld puddle. GTAW produces high-quality welds without slag or oxidation. Because there is no flux, there is no corrosion due to slag entrapment. As a result, minimal postweld cleaning is needed.

A major use of GTAW is to make high-quality welds on aluminum piping. The GTAW process allows greater control of root penetration and fill than almost any other process. For this reason, GTAW is often used to make the root pass on aluminum pipe, even when the remaining passes are made with gas metal arc welding (GMAW), which has a higher filler metal deposition rate.

The V-groove and U-groove welds are typically used for joining medium- and thick-walled pipe used in various piping systems. Piping containing or carrying material that can cause injury to people or damage the environment, or those that operate at elevated pressures, may be categorized as critical piping. Injury or damage occurs when the piping system or other system components fail to contain the material as designed. Welds in critical piping may be required to meet more stringent requirements. Piping that is used for hydronic heating and air conditioning applications, simple water systems, and similar applications may be assessed using less stringent requirements.

This module provides guidance in the application of U-groove welds on aluminum pipe in the 2G, 5G, and 6G welding positions. The dimensions and specifications in this module are representative of codes in general and may not be specific to any particular code. Always follow the proper codes for your site.

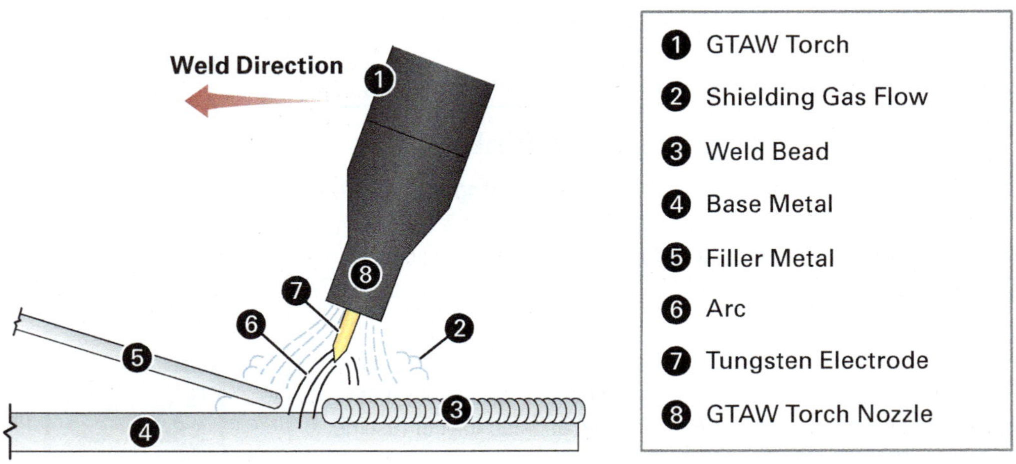

Figure 1 GTAW process.

1.1.0 Safety Practices

Safety considerations are a critical part of preparing for a welding operation. The following is a summary of safety procedures and practices that must be observed while cutting or welding. Before continuing this module, it is strongly recommended that you complete NCCER Module 29101, *Welding Safety*, which provides a thorough review of welding hazards and precautions. Above all, be sure to wear appropriate personal protective equipment (PPE) at all times when welding or cutting.

WARNING!

Welding aluminum is associated with immediate and long-term hazards and health risks that can be reduced and controlled, if not eliminated, by following safety practices. Short-term exposure to welding fumes may cause eye and throat irritation, dizziness, and nausea. Prolonged exposure to welding fumes may cause ulcers, damage to the lungs, kidneys, and nervous system, as well as various types of cancer, including lung, larynx, and urinary tract.

Due to health risks associated with the welding of aluminum, the Occupational Safety and Health Administration (OSHA) has established permissible exposure limits (PELs) for aluminum as well as iron or mild steel. OSHA's PEL limit of 5 mg/m^3 represents the total concentration of fume as an exposure limit over an eight-hour time weighted average. Welders adhere to PELs in addition to following safety practices and protocols of protective clothing, fire prevention, fall prevention, and venting.

1.1.1 Protective Clothing and Equipment

Protective clothing and equipment are a necessary part of every welding task (*Figure* 2). The following general guidelines review typical PPE required for welding:

- *Welding hood* — A welding hood equipped with a properly shaded lens is best for all forms of welding. The welding hood lens must be an appropriate shade for the application.
- *Clothing and gloves* — Wear proper leather or flame-retardant clothing along with welding gloves for protection from flying sparks, molten metal, and heat.
- *Footwear* — Wear high-top safety shoes or boots. Make sure the tongue and lace area of the footwear is covered by a pant leg. If the tongue and lace area is exposed or if the footwear must be protected from burn marks, wear spats under the pant leg or leather chaps over the front and top of the footwear.
- *Caps and hard hats* — Wear a 100% cotton cap without any mesh material. If a hard hat is required for the environment, use one that allows the attachment of rear deflector material and a face shield. A hard hat with a rear deflector may be required when working overhead and may be required by some employers and jobsites.
- *Ear protection* — Wear earmuffs, or at least earplugs, to prevent sparks from entering the ear canals.

Figure 2 GTAW welder with the proper PPE.
Source: © Miller Electric Mfg. LLC

Welders must also familiarize themselves with any unique requirements an employer or job may have regarding PPE.

WARNING!

Welding aluminum is associated with specific hazards and risks that require awareness and safety protocols. Reflected ultraviolet (UV) light is especially challenging when welding aluminum. Aluminum's highly reflective surface poses a big concern. When welding aluminum, ensure all exposed skin is covered by wearing proper PPE and using welding screens to protect others. Additionally, it is important to follow all other OSHA safety precautions.

1.1.2 Fall Protection and Housekeeping

Welding in the construction industry occurs in a variety of different environments. Welding inside large, open shops exposes a welder to equipment hazards on the floor and overhead. Welding outdoors exposes the welder and the welding job to potential weather hazards. In both environments, welders often must crawl under or reach over objects to complete their work.

At times, welders working indoors and outdoors must work at height. Fall protection equipment is required by OSHA when working at a height of 6' or higher (*Figure 3*).

Fall protection equipment saves lives. Always keep fall protection devices untangled and properly attached. Combined with the other equipment a welder must carry, there is a great deal to manage, but fall protection equipment must always be the priority. Never attach any kind of welding gear to the harness that could interfere with its ability to perform its intended function.

Regardless of where you work, always keep the work area as neat as possible. This makes the area safer and helps reduce distractions. Mindful actions, such as making sure that welding cables are not tripping hazards, contribute to a safer environment. Electrode stubs and unused electrodes must not be left on the floor. They can cause someone to lose their footing and fall. Parts waiting to be welded or left over from previous work are a tripping hazard and should be kept out of the walking area. If the parts have sharp edges, they could cut welding cable insulation or torch gas hoses.

Friction Stir Welding

The marriage between steel and aluminum has long been a problem in the welding world due to their extreme differences in thermomechanical properties. Fusion welding of these materials tends to form compounds that are too brittle for reliability. Friction stir welding (FSW) is changing that, and the process is equally effective with aluminum alone.

FSW works by applying a tool to the surface of the two metals along the desired joint. The process is autogenous (requires no filler metal). Through friction alone, enough heat is generated to join the two metals, although they never fully melt. The process has already been used in a few specialized applications, such as the front subframe of the 2013 Honda Accord and the external fuel tank on the space shuttle. Boeing presently uses the process in aircraft structures, and automotive wheel manufacturers are making use of it. The process creates less distortion, doesn't produce a significant heat-affected zone (HAZ), and suffers no loss of strength associated with the use of a filler metal.

Source: Image copyright TWI Ltd, reproduced by permission

Figure 3 Working at height.
Source: Rich Samanich, M.Ed., SCWI

1.1.3 Fire/Explosion Prevention

Flammable gases are commonly found around welding and cutting operations, as are sparks and other significant sources of ignition. For that reason, and because many fire-related incidents over the years have been traced back to welding operations, it is crucial that fire and explosion prevention be given consistent attention. Follow these guidelines to maintain a safe environment:

- Never carry matches, gas-filled lighters, or battery powered devices such as cell phones or vape pens. Sparks can cause the matches or lighters to ignite, or the battery packs to overheat and explode. Any ignition or explosion can cause serious personal injury.
- Never use oxygen to blow dust or dirt from clothing. Traces of the oxygen can remain trapped in the fabric for some time. If a spark hits the fabric during this time, the clothing can burn rapidly and violently out of control.
- Make sure any flammable material in the work area is removed or shielded by a fire-resistant covering. Approved fire extinguishers must be available before attempting any heating, welding, or cutting operations. If a hot work permit and a fire watch are required, be sure these are in place before beginning.
- Ensure all site safety requirements are being observed.
- Never release a large amount of oxygen or use oxygen in place of compressed air. Its presence around flammable materials or sparks can cause rapid and uncontrolled combustion. Keep oxygen away from oil, grease, and other petroleum products.
- Never release a large amount of fuel gas, especially acetylene. Heavier than air, propane tends to concentrate in and along low areas and can ignite at a considerable distance from the release point. Acetylene is lighter than air but is even more dangerous than propane. When mixed with air or oxygen, acetylene will ignite at much lower concentrations than any other fuel gas.
- Even though shielding gases are not toxic, any gas released in a sufficient quantity can displace the oxygen in a small room or enclosed work area, creating an atmosphere with little or no oxygen. Avoid large releases of any type of gas, regardless of its characteristics.
- To prevent fires, maintain a neat and clean work area, and make sure that any metal scrap or slag is cold before disposal.
- Before cutting containers such as tanks or barrels, check to see if they have contained any explosive, hazardous, or flammable materials, including petroleum products, citrus products, or chemicals that decompose into toxic fumes when heated. As a standard practice, always clean and then fill any tanks or barrels with water or purge them with a flow of inert gas, such as nitrogen or argon, to displace any oxygen. Containers must be cleaned by steam cleaning, flushing with water, or washing with detergent until all traces of the material have been removed. Proper procedures for cutting or welding hazardous containers are described in the *American Welding Society (AWS) F4.1, Safe Practices for the Preparation of Containers and Piping for Welding, Cutting, and Allied Processes.*

WARNING!

Welding or cutting must never be performed on drums, barrels, tanks, vessels, or other containers until they have been emptied and cleaned thoroughly, filled with water, or purged with inert gas. All flammable materials and all substances (such as detergents, solvents, greases, tars, or acids) that might produce flammable, toxic, or explosive vapors when heated need to be removed or stable to create a safe atmosphere for welding. Clean containers only in well-ventilated areas. Vapors can accumulate during cleaning, causing explosions and serious personal injury.

1.1.4 Work Area Ventilation

Many gases are introduced to the atmosphere through welding processes. Some of these gases are harmful, but all of these gases can displace or combine with atmospheric oxygen in a confined space. In the case of aluminum, GTAW applies high heat to melt the aluminum and its alloys which facilitates oxidation. Aluminum oxide particles are formed, which mix with the welding fumes. The welding fumes produce white smoke containing the known lung irritants, magnesium oxide and aluminum oxide.

Follow these guidelines, as well as those of your employer, to ensure proper work area ventilation:

- Follow all confined space procedures before welding or cutting in a confined space. Use powered fume extraction systems, as needed.
- Never use oxygen to provide ventilation in confined spaces.
- Always perform cutting or welding operations in a well-ventilated area.

The welding of aluminum is associated with specific gases and fumes including ozone, nitrous oxides, carbon monoxide, and carbon dioxide. For example, ozone is a known lung irritant, so proper ventilation needs to be utilized to minimize the risk. Ozone is produced during GTAW of aluminum. The UV rays from the welding arc interact with oxygen in the area forming ozone gas.

The composition and quantity of the gases and fumes produced from welding aluminum are dependent upon the welding process, filler wire used, coatings on the metal being welded (i.e. paint, anodizing, or Alodine coatings), number of welders, volume of the work area, quality and the amount of ventilation, position of the welder with respect to the fume plume, and presence of contaminants in the atmosphere such as chlorinated hydrocarbon vapors from cleaning and degreasing activities.

For long-term cutting or welding of aluminum and other materials producing toxic gases and fumes, always wear an approved full-face, supplied-air respirator that uses breathing air supplied from outside the work area. For occasional short-term exposure, a HEPA-rated or metal fume filter may be used on a standard respirator.

In addition to controlling work area ventilation for aluminum-specific risks, powered fume extraction equipment must be used when welding materials that contain zinc, cadmium, hexavalent chromium, manganese, and similar materials that produce hazardous fumes if the area is not naturally well-ventilated. *Figure 4* shows an example of a powered air purifying respirator (PAPR).

WARNING!

Breathing in fumes from welding aluminum can cause metal fume fever, also called *welder's flu*. Symptoms include chest tightness, cough, metallic taste, fever, and headache. Proper ventilation on the worksite and individual PPE are necessary to limit exposure and potential risk from fumes created by welding aluminum.

1.2.0 GTAW Techniques

GTAW weld bead characteristics and quality are affected by several factors, each of which is influenced by the way the welder handles the torch. These factors include the following:

- Torch travel speed and arc length
- Torch angles
- Torch and filler metal handling techniques
- Consistent feeding of the filler metal

Weld Puddle Size

When performing GTAW in the horizontal, vertical, and overhead positions, you must keep the weld puddle smaller than it is in the flat position. The weld puddle is generally held in the joint by surface tension. When the puddle is too large, gravitational force pulling down on the puddle overcomes surface tension and the liquid metal runs out of the joint.

Figure 4 Powered air purifying respirator (PAPR).
Source: Rich Samanich, M.Ed., SCWI

1.2.1 Torch Travel Speed and Arc Length

Torch travel speed and arc length affect the GTAW weld puddle and weld penetration. Slower travel speeds allow more heat to concentrate and form larger, more deeply penetrating puddles. Faster travel speeds prevent heat buildup and form smaller, shallower puddles.

Arc length is the major control for bead width. As the torch is raised, arc length, voltage, and bead width increase.

1.2.2 Torch Angles

The two basic torch angles that must be controlled when performing GTAW are the work angle and the travel angle. The definition of these angles is the same as for all other types of arc welding.

The torch work angle (*Figure 5*) is a 90° or lesser angle between a line perpendicular to the major workpiece surface at the point of electrode contact and a plane determined by the electrode axis and weld axis. For a T-joint or corner joint, the line is perpendicular to the non-butting member. For pipe, the plane is determined by the electrode axis and a line tangent to the pipe surface at the same point. As shown in *Figure 5*, when the torch is held perpendicular, it is referred to as a *0° work angle*. To properly position beads in a groove joint, the work angle may be changed 20° in either direction.

The torch travel angle is a 90° or lesser angle between the electrode axis and a line perpendicular to the weld axis at the point of electrode contact in a plane determined by the electrode axis and weld axis. For pipe, the plane is determined by the electrode axis and a line tangent to the pipe's surface at the same point. The travel angle used for GTAW is a push angle (*Figure 6*). The push angle is created when the torch is tilted back, away from the direction of travel, so that the electrode tip precedes the torch in the direction of the weld. In this position, the electrode tip and shielding gas are directed just ahead of the weld bead.

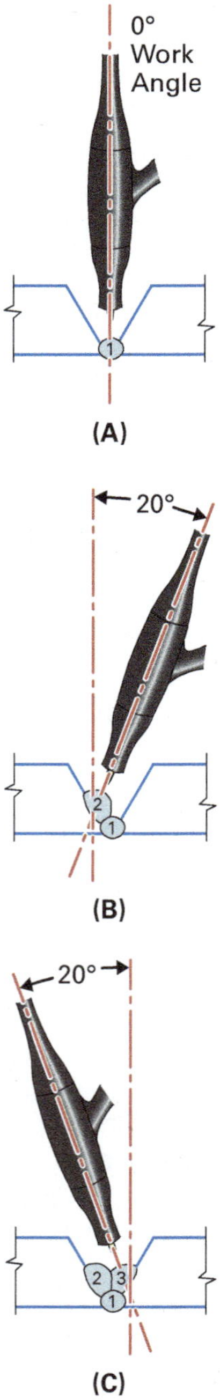

Figure 5 Typical GTAW torch work angles.

This provides a cleaning action on the positive portion of the alternating current (AC) voltage cycle. Push angles of 15° to 20° are normally used for GTAW on aluminum.

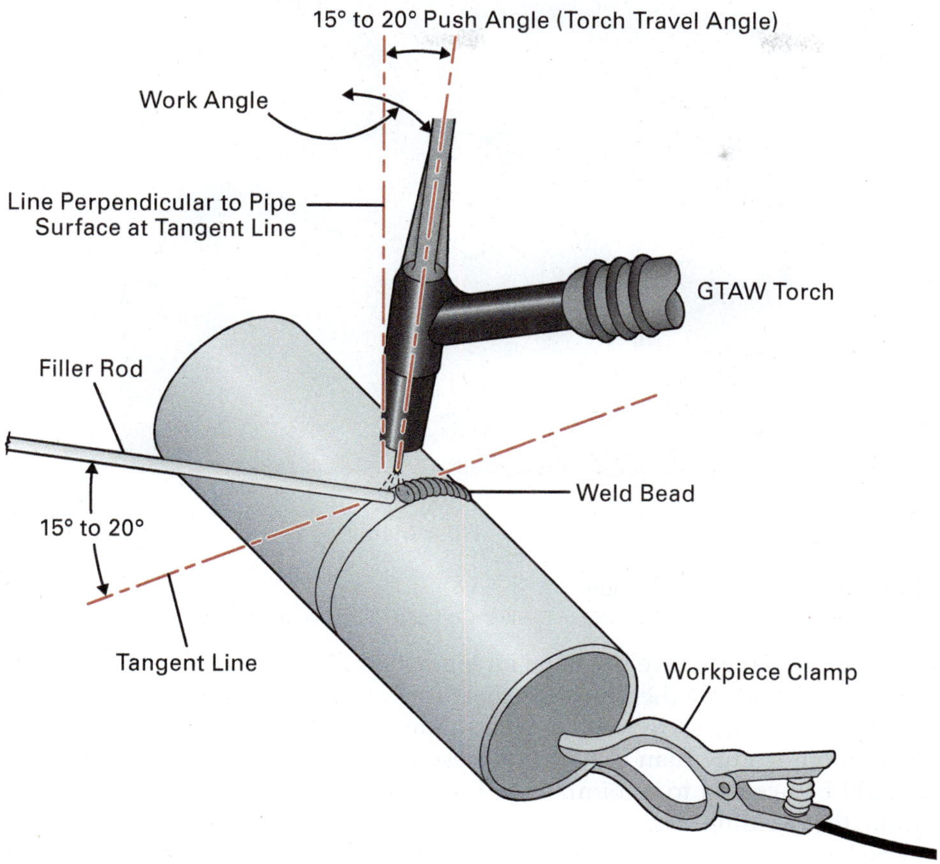

Figure 6 GTAW torch travel angles.

1.2.3 Torch and Filler Metal Handling Techniques

The freehand technique is the preferred torch and filler metal handling method for performing GTAW on aluminum.

In this method, the torch electrode tip is held just above the weld puddle or base metal (*Figure 7*). The torch is supported by the welder's hand like a pencil. The hand is usually steadied by resting some part of it on or against the base metal. This helps to keep the proper arc length (distance from the electrode tip to the weld puddle). Due to the high thermal conductivity of aluminum, it may help to insert a piece of nonflammable insulation between the gloved hand and the workpiece. As needed, the welder can move the torch tip in a small circular motion within the molten puddle. This maintains the puddle size and advances the puddle. Filler metal is added as needed.

The GTAW filler metal rod is held in the hand that is not supporting the torch. Hold the rod at an angle of 15° to 20° above the base metal surface and in line with the weld. When a 15° to 20° push angle is also used, the rod remains roughly perpendicular to the electrode axis. Always keep the tip of the rod within the shielding gas envelope to protect it from atmospheric contamination and to keep it preheated.

Insert the rod into the leading edge of the weld puddle, using extreme care not to touch the tungsten electrode to the end of the rod. If the tungsten electrode touches the rod or weld puddle, the electrode will become contaminated with filler metal. The electrode must then be removed, the contaminated portion broken off, and the end reformed before welding can continue. To prevent contamination of the electrode by the filler metal, move the electrode to the back edge of the weld puddle as the rod is inserted into the weld puddle's leading edge. Timing those two actions helps prevent tungsten contamination.

CAUTION

Excessive push angles tend to cause air to be drawn into the weld from under the back edge of the torch nozzle. The air mixes with the shielding gas stream and contaminates the weld.

Figure 7 Freehand technique.
Source: Photo courtesy of The Lincoln Electric Company, Cleveland, OH, U.S.A.

When making weave beads on strengthened aluminums (i.e. -T4, -T5, -T6, or -H conditions), whether they are heat-treated or strain-hardened, they lose strength when they are welded. Some aluminum alloys lose strength and ductility more readily than others. For these reasons, weave beads on aluminum should be avoided to minimize the loss of strength caused by the heat input from the welding arc. From a practical standpoint, only aluminum in the annealed condition can be welded without experiencing a degradation of mechanical properties.

1.2.4 Consistent Feeding of the Filler Metal

Feeding the filler metal consistently as the weld progresses is important to ensure the weld puddle develops evenly and produces an even bead. Like all welding techniques, the technique described here requires practice to master but is well worth the effort.

Cradle the rod between the pointer finger and the middle finger, similar to the way you might hold a pencil. Then use the thumb to consistently push the rod forward in small motions. Once the thumb is extended pushing the wire forward, quickly bend and reposition the thumb to begin a new forward push. Consistent motion is important, so the thumb must be repositioned quickly to maintain the pace. The thumb can also be used to slowly roll the rod as well during the pushing motion. This technique is easily practiced without actively welding; simply practice with a rod at home or during free time to become comfortable with the technique before trying it on a weld.

Another technique to consider is referred to as the *finger-pull technique*. The rod is secured between the pointer and middle fingers, which are used to pull the electrode forward. At the end of the forward stroke, the fingers are slipped backward along the rod to reposition, without pulling it backward and out of the weld puddle in the process. Once the two fingers are repositioned, the pulling motion begins again.

Heat Protection

GTAW welding often places the welder's hand in direct contact with the workpiece to steady the torch and maintain a consistent arc length. However, especially when welding aluminum, the workpiece can become extremely hot. Even the best gloves are not enough to effectively protect the welder's hand, leading to the possibility of injury and more weld bead terminations due to discomfort.

There are several products available that can provide extra protection against heat as well as a surface that slides across the workpiece easily without being too clumsy and interfering with the welder's dexterity. Some are designed to fit over a portion of the hand or multiple fingers, while other products slide over a single finger. One thing is certain—a lot of GTAW welders are thankful for the added protections that allow the welding of very long beads without burns or discomfort.

Source: Griffin Heat Shield Co., www.griffinheatshield.com

1.3.0 Preparations for Welding

Before welding practice can take place, the work area and welding equipment must be prepared, and the coupons to be welded must be cut and cleaned. The following sections explain how to prepare for GTAW of aluminum pipe.

To practice welding, you will need a welding table, bench, or stand. The preferred welding surface is aluminum. Provisions must be available for mounting practice welding coupons out of position.

To set up the area for welding, follow these steps:

Step 1 Make sure the area is properly ventilated. Use doors, windows, fans, and localized fume extraction equipment as needed.

Step 2 Check the area for fire hazards. Remove any flammable materials before proceeding.

Step 3 Locate the nearest fire extinguisher. Do not proceed unless the extinguisher is charged and you know how to use it.

Step 4 Set up welding screens around the welding area.

1.3.1 Aluminum Filler Metals

Filler metals are selected to be compatible with the base metal to be welded. The Welding Procedure Specification (WPS) or site standards will specify the proper filler metal type and size to use.

Unlike steel, the aluminum filler metal will not develop a tensile strength equal to that of the base metal. As a result, aluminum weld joints are the weakest link in a welded assembly. This important fact is another reason why proper filler metal selection for aluminum pipe is critical.

For all welding exercises associated with this module, weld the aluminum pipe coupons using $^{3}/_{32}$" and/or $^{1}/_{8}$" aluminum filler metal rods as specified by your instructor. Remove only a small number of filler rods at a time. Keep the remainder in the package to keep them clean. Before using the filler metal, check it for burned ends or contamination, such as oxidation, dirt, oil, or grease, all of which can cause weld discontinuities. Clean the filler metal with a clean rag and acetone. Use emery cloth or stainless steel wool to remove oxides if necessary and snip any burned ends. If the filler metal cannot be cleaned effectively, do not use it.

CAUTION

Always use filler metal as specified by the WPS or site quality standards, if applicable, to prevent weld discontinuities and satisfy the requirements of the weld.

Going Green

Conserve Aluminum Materials

Aluminum is expensive and difficult to obtain for practice purposes. To conserve resources and reduce waste, completely use all weld coupons until all surfaces have been welded upon. Cutting coupons apart and reusing the pieces conserves materials. Use flat material that cannot be cut into weld coupons to practice running beads. Note that stringer beads can be practiced while leaving enough room between them to practice weave beads as well. Any aluminum material that is no longer suitable for welding practice should be recycled.

1.3.2 Practice Pipe Weld Coupons

Pipe weld coupons should be cut from Schedule 40 aluminum pipe that is 3" to 12" (DN80 to DN300) in nominal diameter. Each welded joint requires two coupons of the same size. *Figure 8* shows typical AWS aluminum pipe specifications for V-groove and U-groove joint preparation. The U-groove joint is preferred when there is no backing.

Welding practice for this module should be done using the U-groove joint shown in *Figure 8*. The V-groove joint shown can be used, although a backing gas is needed. A backing gas is not required for the U-groove joint.

Aluminum is easily contaminated. Proper coupon preparation is very important. Aluminum pipe coupons must be cut to the desired lengths, cleaned, and beveled to the proper angles before they are positioned for welding. The preferred technique for beveling and/or adding a root face is mechanical machining. Grinding is also acceptable, although machining the grooves and root faces of aluminum pipe is the cleanest and most uniform way to prepare pipe for GTAW. Thermal cutting techniques are not considered acceptable. If thermal cutting techniques must be used to cut aluminum alloys, at least $^{1}/_{8}$" (3.2 mm) of the cut surface must be removed by machining or grinding before welding.

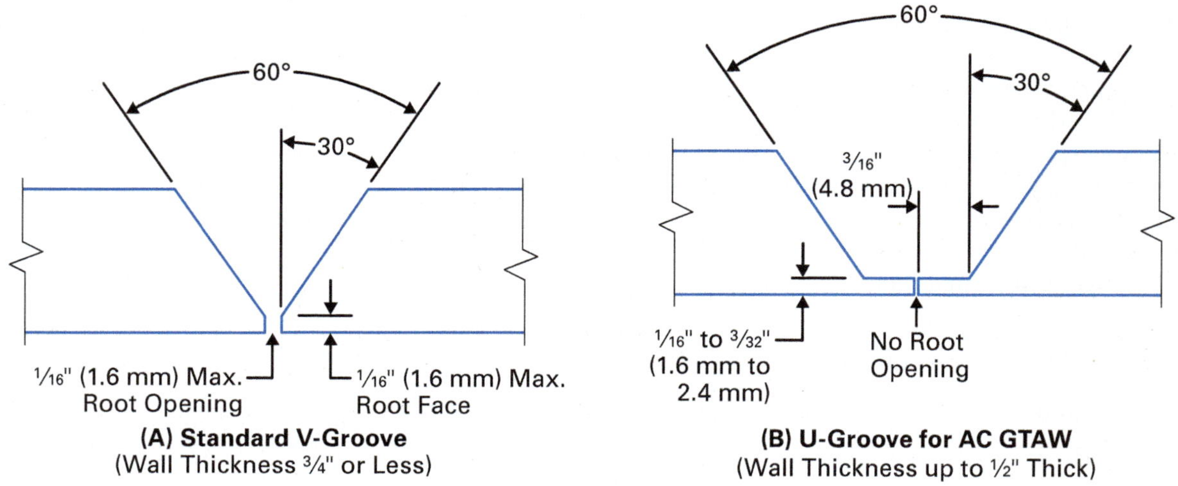

Figure 8 AWS aluminum pipe groove specifications for use without backing.

CAUTION

Grinding wheels, files, and similar tools used on aluminum should not be used on any other type of metal or on surfaces contaminated with grease or oil.

NOTE

When using the U-groove joint, it is best to clean the backside of the weld (the inside of the pipe) with acetone or other approved solvent to remove surface contaminants. Avoid cleaning with wire brushes or grinders. The high melting point of the thin film of oxidation left behind can help provide a form of backside protection to the joint.

WARNING!

Grinding wheels used on aluminum must be made for aluminum only. Using other types of grinding wheels on aluminum may cause the grinding wheels to shatter.

Clean oil and grease from the coupons with an approved substance such as acetone. After oil and grease have been removed, use a stainless steel wire brush to remove all other contaminants and oxidation. Use only light pressure on the brush.

WARNING!

To prevent injury when using cleaning solvents, always comply with the Safety Data Sheet (SDS) guidelines as well as your company's guidelines for the solvents being used.

To prepare pipe coupons for groove weld joints, follow these steps:

Step 1 Bevel or groove the end of the pipe, as shown in *Figure 8*, using any acceptable method discussed in *Section 1.3.2*.

Step 2 Check the groove angle and root face for accuracy. There should be no dross. The angle and dimensions should be as shown in *Figure 8*, with no notches deeper than $^1/_{16}$" (1.6 mm). Note, if using a thermal cutting technique, follow proper procedures, as stated in *Section 1.3.2*.

Step 3 Cut off the section with the beveled pipe end. The piece should be a minimum of 4" (10 cm) long.

Step 4 Prepare a second matching section of pipe to complete the coupon assembly.

Step 5 Use an approved solvent to remove any grease or oil within 1" (2.5 cm) of the weld joint. Clean the outside of the aluminum pipe with a grinder or stainless steel brush as needed. As noted above, do not use a grinder or wire brush on the inside of the pipe when using the U-groove joint; the layer of oxidation inside can help provide some backing protection.

CAUTION

To prevent contamination, do not use a stainless steel wire brush on anything else after it has been used on aluminum.

Step 6 Align the two pipe coupon sections so their inside surfaces are even all around. The coupons must also be aligned so the resulting welded pipe assembly is straight. Align small-diameter pipe by clamping both pieces to a piece of angle iron. Larger pipe can be aligned with the aid of a pipe alignment jig (*Figure 9*), a Hi-Lo gauge (*Figure 10*) or by holding a straightedge across the joint, parallel to the pipe axis. The straightedge must be used on all four quadrants of the pipe in case one or both sections are distorted.

Step 7 When the pipe ends are properly aligned with no root opening, make the first of four tack welds.

Step 8 Make the second tack weld on the opposite side from the first, making sure there is no root opening.

Step 9 Check the root again and weld the third tack midway between the first two tacks.

Step 10 Weld the fourth tack opposite the third tack and midway between the first and second tacks. There should now be four tack welds evenly spaced at 90° apart around the pipe coupon (*Figure 11*).

Step 11 Feather the tack welds with the edge of a grinding wheel. Feathering the ends of the tack welds with a grinder helps to fuse the root pass into the tack weld.

Welding Large-Diameter Pipe

The practice coupons on large-diameter pipe should be greater than 6" in length. They need more metal to help absorb the increased heat used to make these welds.

CAUTION

When welding pipe, avoid making arc strikes outside of the weld groove on the surface of the pipe. An arc strike can cause a hardened spot that can crack as the pipe expands and contracts. An arc strike on the pipe surface is considered a defect, which will require repair or reworking.

Figure 9 Pipe alignment jig.
Source: Courtesy of Ashley Applegate

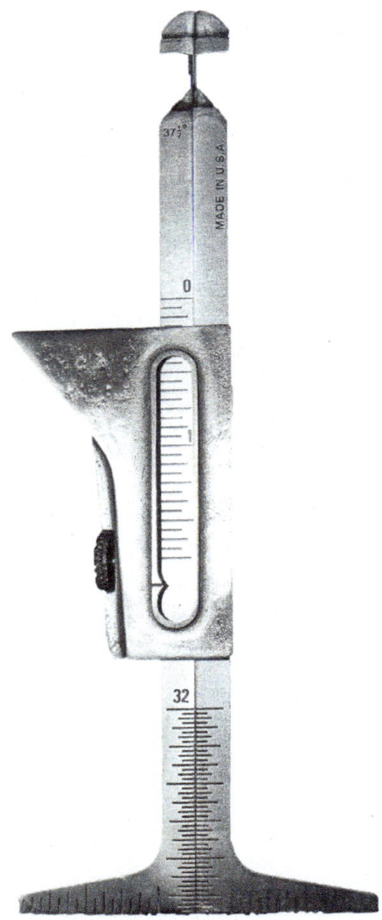

Figure 10 Hi-Lo gauge.
Source: Rich Samanich, M.Ed., SCWI

Aluminum Pipe Weld Backing

Open-root and consumable insert welding do not generally work well with aluminum piping. When you need to make an open-root groove weld, the welder must back it with a temporary or permanent backing ring; otherwise, the root opening must be eliminated from the standard V-groove joint configuration. No gas backing is required when the U-groove joint is used or when a temporary or permanent backing ring is used with a V-groove joint. The U-groove joint is recommended for manual AC GTAW of pipe.

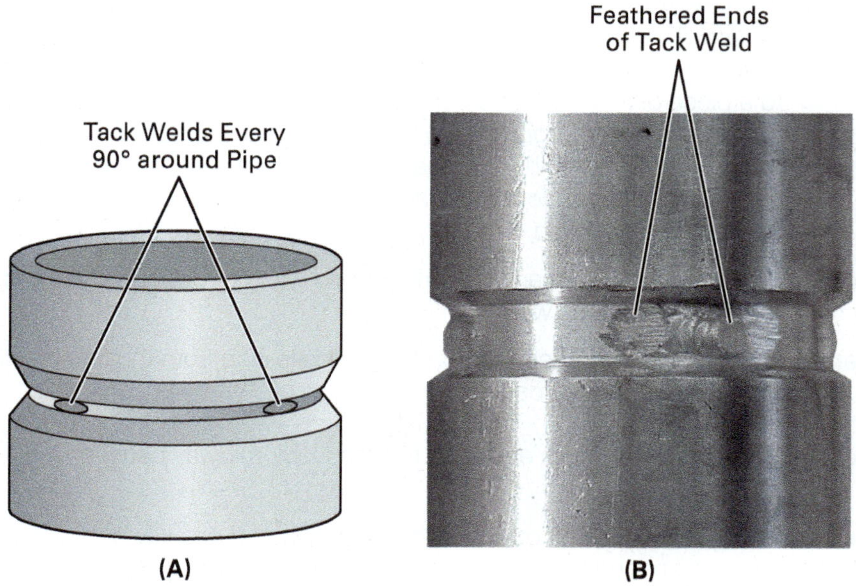

Figure 11 U-groove weld coupons tacked together.
Source: Terry Lowe (11B)

1.4.0 Welding Equipment

Welding aluminum with GTAW typically requires equipment that provides AC with an argon shielding gas. Aluminum GTAW requires a significant amount of welding current due to its thermal conductivity; heat rapidly flows away from the weld joint. Ensure that the equipment selected is capable of providing sufficient current at a reasonable duty cycle. Welding aluminum at welding current levels less than those required means that the weld cannot progress at the proper speed. As a result, the workpiece will overheat and create an inferior weld. Because of the higher currents associated with aluminum welding, water-cooled torches are recommended. Of course, this will not reduce the temperature of the workpiece itself.

Conventional AC machines used for GTAW are typically based on sine wave or square wave technology. Newer inverter technology units have additional features, such as high-frequency start only and current sequencing (start/stop ramping), that help control puddle formation at the beginning of the weld and in filling the termination crater. Note, scratch start should not be used on aluminum as it can leave tungsten inclusions. The newer inverter technology machines provide additional control over important variables that affect the quality of aluminum welds. The ability to vary both the AC wave balance and frequency is preferred for aluminum welding. Control over the balance between the electrode negative (EN) and electrode positive (EP) portions of the AC sine wave allow the operator to control two important functions:

- As the wave is unbalanced to the EN side, the current-carrying capacity of the electrode is increased and the amount of oxide removal, also called *cleaning action*, is decreased.
- When welding on new material that is properly cleaned, the wave can be set at 80% EN or more. This allows the use of a smaller electrode at higher welding currents.

Conventional welding machines in the United States use 60-cycle electrical power as supplied by the line current. (Many other countries use 50 cycles.) Newer inverter machines are able to vary the frequency, potentially increasing it to 300 cycles per second or even more. This eliminates the need to use balled pure- or zirconiated-tungsten electrodes and permits the use of a truncated EWTh2 or other tungsten electrodes. Increased cycles per second also provides better arc control and permits smaller weld sizes. Frequencies of 120 cycles to 140 cycles are typically used to weld aluminum.

Although pure tungsten (EWP) electrodes are used for welding aluminum with AC and high-frequency machines, zirconiated-tungsten electrodes (EWZr) are often preferred due to their higher current-carrying capacity and lower susceptibility to contamination. They also hold a balled tip better when welding aluminum.

Table 1 shows the current ranges for various electrode types and sizes.

NOTE

Thoriated tungsten electrodes should not be used to weld aluminum. These electrodes tend to spit tungsten into the weld pool when used on AC. Professional welding societies and manufacturers, including the American Welding Society, the Aluminum Association, Miller Electric, and the Lincoln Electric Arc Welding Foundation, do not recommend thoriated tungsten for AC welding. Ceriated or 2% lanthanated tungsten are the types of tungsten recommended for use.

TABLE 1 Typical Current Ranges for Tungsten Electrodes

Electrode Diameter	Direct Current (Amps)		Alternating Current (Amps)			
	DCEN	DCEP	Unbalanced Wave		Balanced Wave	
	EWCe-2 EWLa-1 EWLa-1.5 EWLa-2 EWP EWTh-1 EWTh-2	EWCe-2 EWLa-1 EWLa-1.5 EWLa-2 EWP EWTh-1 EWTh-2	EWP	EWCe-2 EWLa-1 EWLa-1.5 EWLa-2 EWTh-1 EWTh-2 EWZr	EWP	EWCe-2 EWLa-1 EWLa-1.5 EWLa-2 EWTh-1 EWTh-2 EWZr
0.01" (0.3 mm)	up to 15	*	up to 15	up to 15	up to 15	up to 15
0.02" (0.5 mm)	5 to 29	*	5 to 15	5 to 20	10 to 20	5 to 20
0.04" (1 mm)	15 to 80	*	10 to 60	15 to 80	20 to 30	20 to 60
1/16" (1.6 mm)	70 to 150	10 to 20	50 to 100	70 to 150	30 to 80	60 to 120
3/32" (2.4 mm)	150 to 250	15 to 30	100 to 160	140 to 235	60 to 130	100 to 180
1/8" (3.2 mm)	250 to 400	25 to 40	150 to 210	225 to 325	100 to 180	160 to 250
5/32" (4 mm)	400 to 500	40 to 55	200 to 275	300 to 400	160 to 240	200 to 320
3/16" (4.8 mm)	500 to 750	55 to 80	250 to 350	400 to 500	190 to 300	290 to 390
1/4" (6.4 mm)	750 to 1000	80 to 125	325 to 450	500 to 630	250 to 400	340 to 525

DC and AC Electrode Legend

EWCe-2: Tungsten + 2% Ceria EWLa-1: Tungsten + 1% Lanthana EWLa-1.5: Tungsten + 1.5% Lanthana EWLa-2: Tungsten + 2% Lanthana	EWP: Pure Tungsten EWTh-1: Tungsten + 1% Thoria EWTh-2: Tungsten + 2% Thoria EWZr: Tungsten + Zirconia

Notes:
All values are based on the use of argon as the shielding gas. Other current values may be employed depending on the shielding gas, type of equipment, and application.
*These combinations are not commonly used.

Source: Courtesy of The Lincoln Electric Company, Cleveland, OH, U.S.A.

Once an appropriate welding machine has been selected, follow these steps to prepare it:

Step 1 Verify that the welding machine can be used for GTAW. If a water-cooled torch will be used, select an appropriate cooling unit.

Step 2 Select an air-cooled or water-cooled GTAW torch. Make sure it is compatible with the welding machine and any optional cooling unit. Water-cooled torches are preferred. Equip the torch with an appropriate cup size for the technique to be used (freehand recommended).

Step 3 Identify the location of the primary power disconnect switch.

Step 4 Configure the welding machine for aluminum GTAW welding (*Figure 12*) as directed by your instructor. Configure the torch polarity to AC. Equip the torch with a properly prepared $^{3}/_{32}$" or $^{1}/_{8}$" (2.4 mm or 3.2 mm) EWP or EWZr tungsten electrode.

Step 5 Connect the proper argon or helium-argon shielding gas for the application as directed by your instructor, filler metal manufacturer, WPS, or site quality standards.

Step 6 Connect the clamp of the workpiece lead to the workpiece.

Step 7 Turn on the welding machine. Adjust the flowmeter and purge the torch according to the manufacturer's instructions.

Step 8 Set the initial welding current for the application.

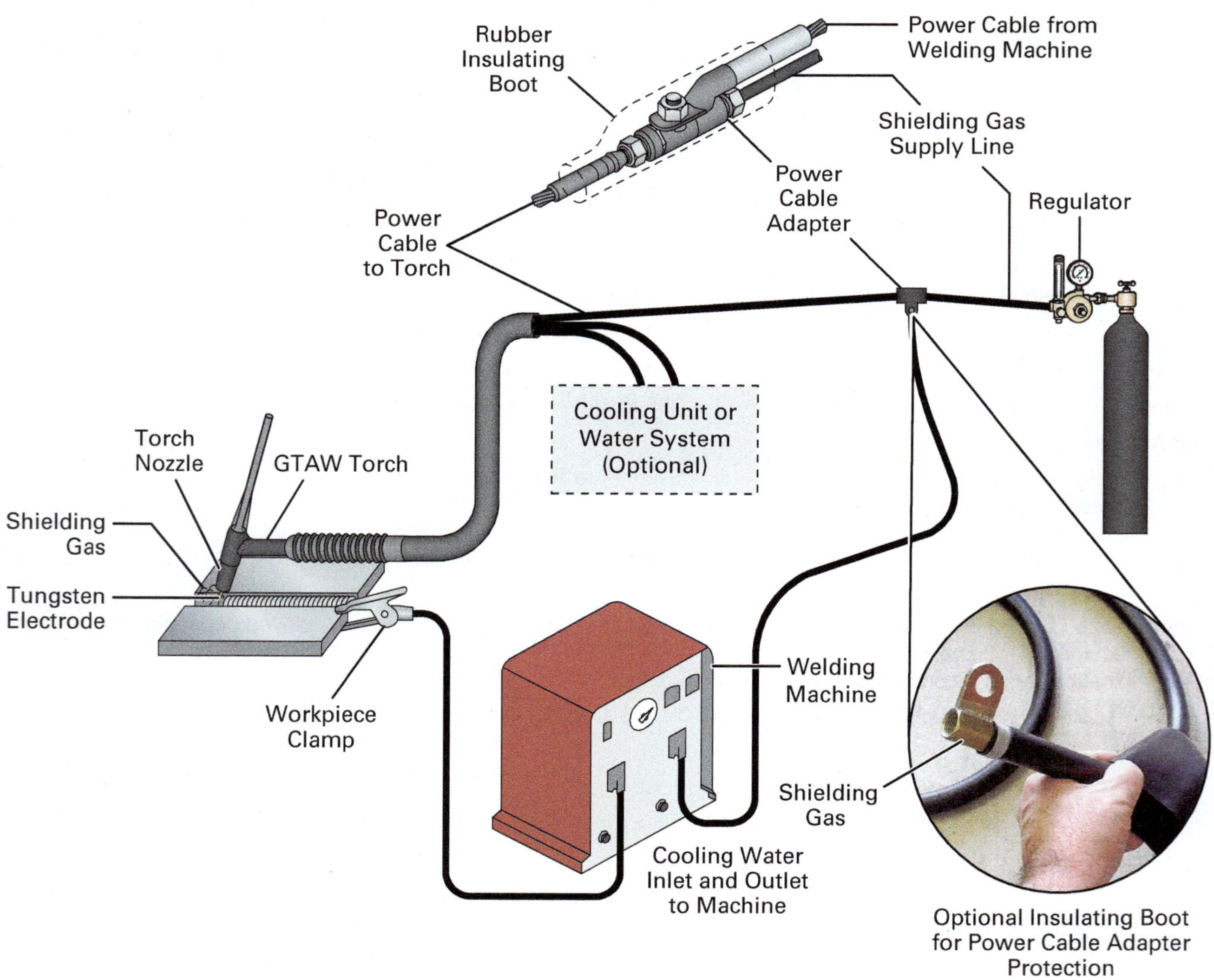

Figure 12 Configuration diagram of typical GTAW welding equipment.

1.4.1 Selecting a Shielding Gas

A shielding gas is selected according to the material being welded. Argon is the most commonly used shielding gas. Argon can be used for welding a wide range of materials, including aluminum, titanium, steel, and stainless steel.

Helium and helium-argon mixtures raise the temperature of the arc. This rise in temperature promotes faster welding speeds and deeper weld penetration. However, helium is expensive and can cause difficulty in starting the arc.

Impurities in the shielding gas or mixture with other gases, such as carbon dioxide (CO_2), will contaminate the weld and change the color of the aluminum (usually to black).

CAUTION

Always use the correct gas when performing GTAW. Accidental use of a gas mix intended for GMAW can cause rapid destruction of the tungsten and contamination of the weld.

Versatile GTAW Machine

The Miller Dynasty 350 TIGRunner is a versatile machine designed to provide superior arc performance with complete control over the arc shape. Auto-Line™ allows the unit to be connected to any voltage worldwide from 208 volts (V) to 575 V. The unit also operates on power supplies using a 50 hertz (Hz) or 60 Hz frequency. Extended AC balance controls the amount of oxide cleaning, which is essential for high-quality welds on aluminum.

Source: © Miller Electric Mfg. LLC

1.5.0 GTAW Techniques and Positions for Aluminum Pipe

Aluminum pipe is commonly welded with AC using argon as a shielding gas. In some cases, a helium-argon mix is used for increased penetration, although adding helium significantly increases material costs. Higher currents are used when welding aluminum because of high thermal conductivity in the base metal. The U-groove joint is preferred for aluminum pipe joints without backing because it reduces heat transfer away from the weld joint. The joint design also allows complete root penetration and fusion with a smaller weld puddle.

A smaller weld puddle is more easily controlled in the vertical and overhead positions by surface tension and molecular attraction than a large puddle. The challenge is to establish a weld puddle that penetrates and fuses with the pipe without falling through to the inside. The most difficult part of welding aluminum is getting the proper penetration and fusion for the root pass when a backing material is not there to prevent the puddle from falling through.

1.5.1 Pipe Groove Weld Test Positions

Groove welds may be made in all positions on pipe. The weld position is determined by the axis of the pipe. Four standard weld test positions are used with pipe (*Figure 13*).

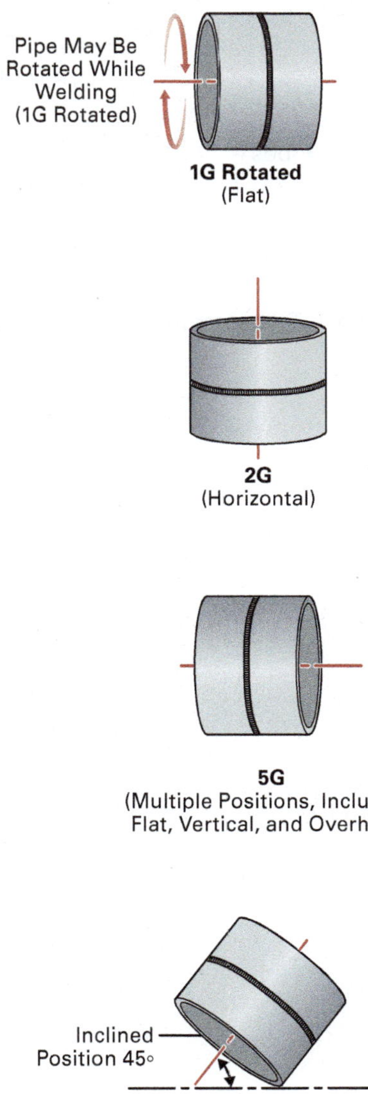

Figure 13 Four basic pipe groove weld test positions.

NOTE

For pipe over 3/8" (9.6 mm) thick, side bends may be taken in lieu of face and root bends. For base metals which exceed 3/4" (19 mm), side bends shall be taken.

To determine the quality of a pipe weld, specific sections of pipe are removed for destructive testing. For positions 1G Rotated and 2G, two specimens are removed. They must be taken from opposite sides of the pipe, as shown in *Figure 14*. To destructively test 5G and 6G welds, the test specimens are cut from the four regions shown in *Figure 14*. Once removed, the specimens are put through different bend tests to determine the quality of the welds.

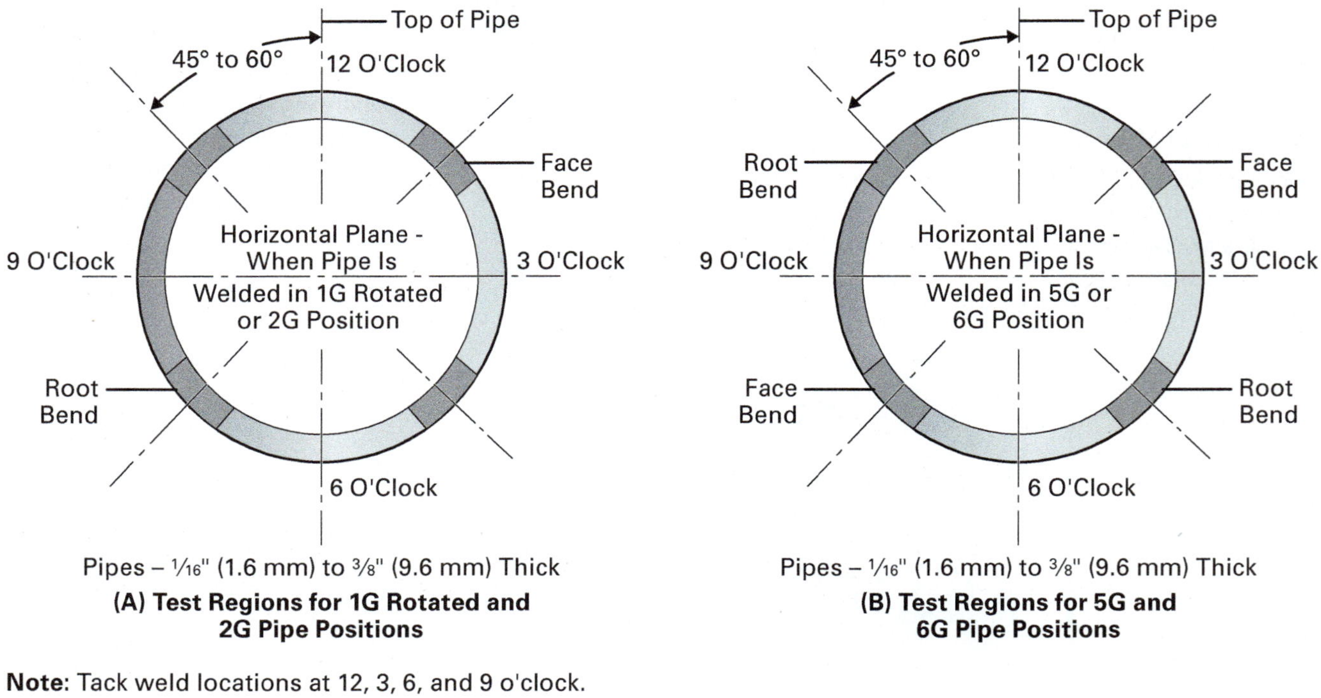

Figure 14 Test specimen regions for pipe positions 1G Rotated, 2G, 5G, and 6G.

1.5.2 Acceptable and Unacceptable Pipe Weld Profiles

Pipe groove welds should be made with slight reinforcement and a gradual transition to the base metal at each toe. The root pass should have complete penetration. The face reinforcement can range from being flush to a maximum of 3/32" (2.4 mm) for materials up to 3/8" (9.6 mm) thick. Excessively large face reinforcement reduces the pipe's strength.

Completed pipe welds should be inspected to ensure they exhibit the following characteristics; if a weld does not meet these criteria, or if it has any of the defects shown in *Figure 15*, it must be repaired:

- Straightness of the bead
- Uniform appearance of the bead face; face and root reinforcement depth in accordance with applicable codes
- Smooth, flat transition with complete fusion at the toes of the weld
- Termination crater filled
- No porosity to exceed 3/32" (2.4 mm)
- Undercut 1/32" (0.8 mm) or less than 10% of base metal thickness
- No overlap
- No overfill
- No cracks

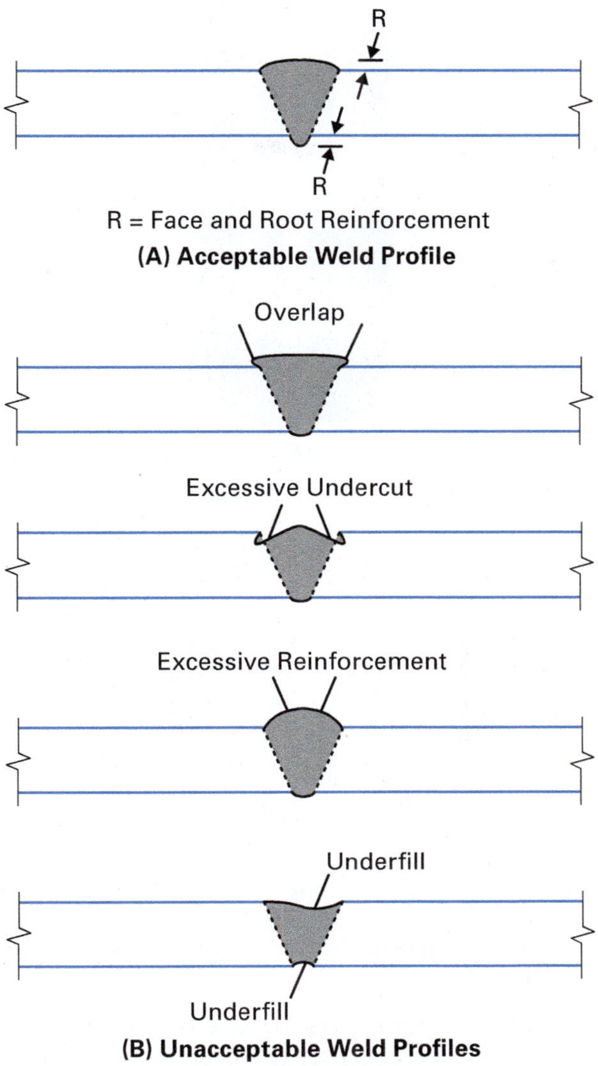

Figure 15 Acceptable and unacceptable pipe groove weld profiles.

1.6.0 Practicing U-Groove Pipe Welds

The following sections explain how to perform U-groove welds in these positions (note that the 1G Rotated position is not practiced in this module):

- Horizontal (2G)
- Multiple (5G)
- Multiple inclined (6G)

1.6.1 Practicing Horizontal (2G) Position Pipe Welds

Practice 2G position pipe welds using the bead sequences and work angles shown in *Figure 15*. Use argon shielding gas and a filler rod of a diameter specified by your instructor. The torch should be held at a 15° to 20° push angle. To better control the thickness of the root reinforcement, the torch work angle can be dropped slightly from the angles shown in *Figure 16*, but not more than 10°. Note that the work angle changes for Passes 2 through 8 to properly place the weld bead. For the root pass, use a slight circular motion to form the weld puddle. Be sure to fuse into the tack welds.

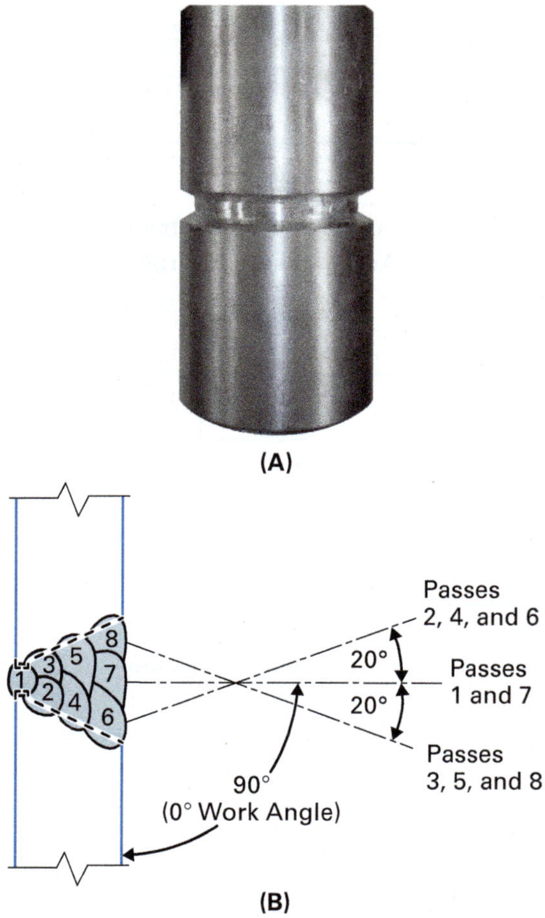

Figure 16 Multipass 2G bead sequences and work angles for the U-groove joint.
Source: Terry Lowe (16A)

When running the remaining passes, use stringer beads and the freehand torch-handling technique. Ensure proper fusion at the toes of the weld bead. Take special care to fill the crater at the termination of the weld.

Follow these steps to practice pipe welds in the 2G position:

Step 1 Tack-weld the practice coupons together, as previously explained.

Step 2 Clamp or tack-weld the coupon into position with the axis of the pipe vertical.

Step 3 Run the root pass (Pass 1), as shown in *Figure 15*, starting on a tack weld. Continue this procedure until the root pass has been completed.

Step 4 Brush the root pass to clean the weld.

Step 5 Run the remaining passes at the appropriate work angles. Overlap the passes and clean each weld pass before beginning the next one.

1.6.2 Practicing Multiple (5G) Position Pipe Welds

Practice the multiple (5G) position pipe welds. Use argon shielding gas and a filler rod of a diameter specified by your instructor. Start on a tack weld positioned at the bottom of the pipe, and weld uphill toward the top. The torch should be held at a 15° to 20° push angle for the root and remaining passes (*Figure 17*). When running the remaining passes, as shown in *Figure 18*, use stringer beads and the freehand torch-handling technique. Ensure proper fusion at the toes of the weld bead. Take special care to fill the crater at the termination of the weld.

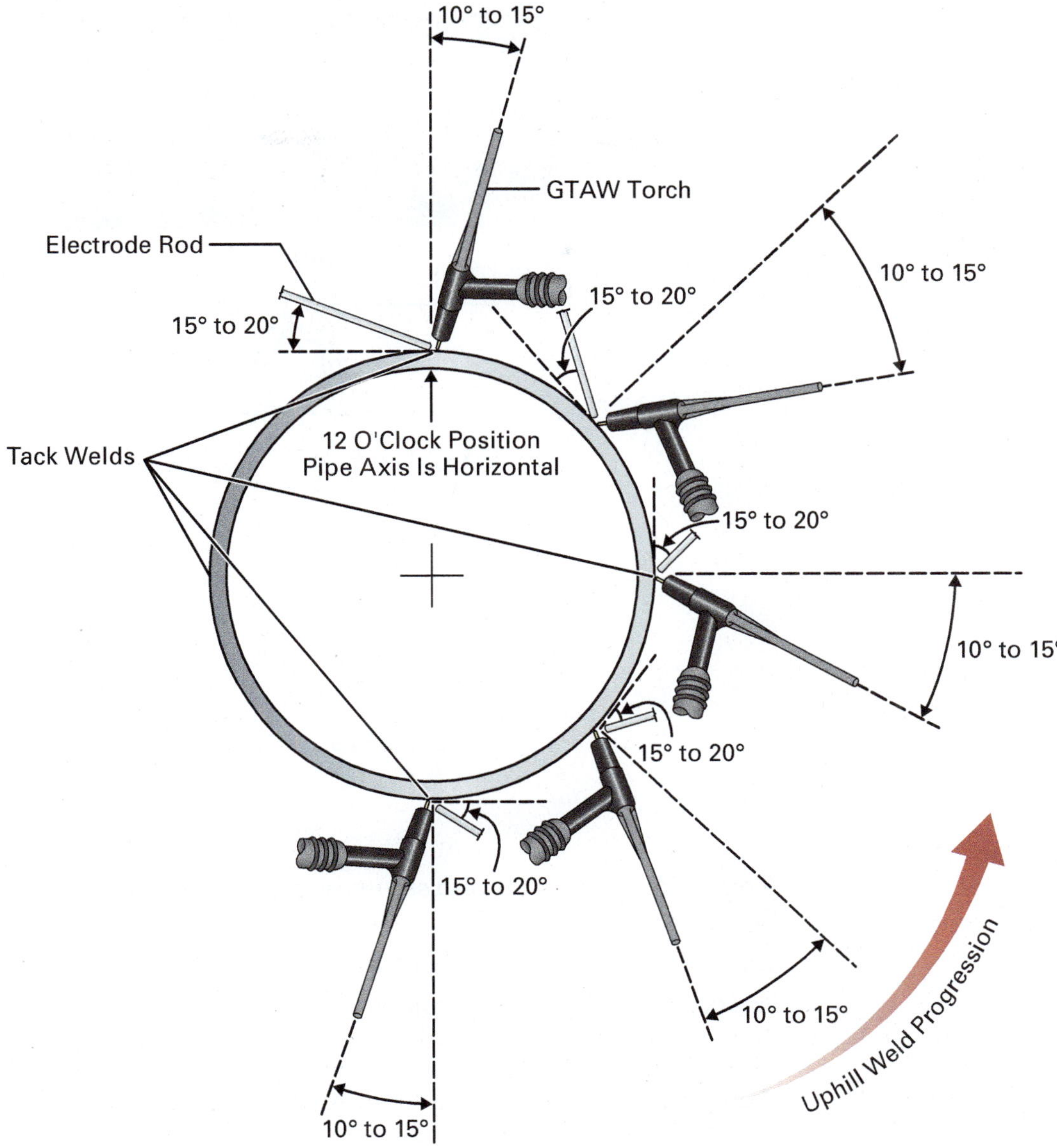

Figure 17 GTAW torch and electrode rod angles for the 5G position.

Follow these steps to practice pipe welds in the 5G position:

Step 1 Tack-weld the practice coupons together, as explained earlier.

Step 2 Clamp or tack-weld the coupon into position with the pipe axis horizontal.

Step 3 Run the root pass uphill with a stringer bead, as shown in *Figure 16*. Repeat the procedure for the opposite side of the pipe.

Step 4 Brush the root pass to clean the weld.

Step 5 Run the remaining passes uphill at the proper angles using the bead sequence shown in *Figure 17*. Overlap the passes and clean the weld after each pass.

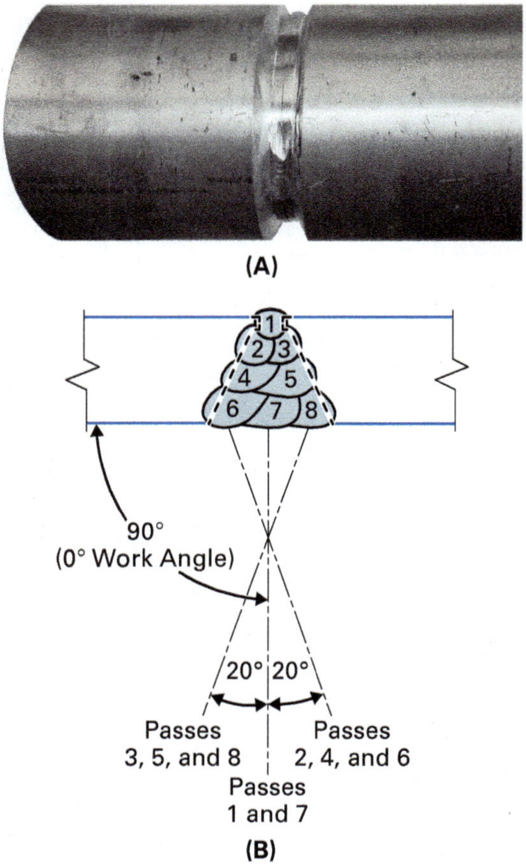

Figure 18 Multipass 5G bead sequences and work angles for the U-groove joint.
Source: Terry Lowe (18A)

1.6.3 Practicing Multiple Inclined (6G) Position Pipe Welds

NOTE

If required for training purposes, a restricting ring may be added to the 6G position coupon to form a 6GR position coupon.

Practice the multiple (6G) position pipe welds. Use argon shielding gas and a filler rod of a diameter specified by your instructor. Start on a tack weld positioned at the bottom of the pipe, and weld uphill toward the top. The torch should be at approximately a 15° to 20° push angle for the root and remaining passes (*Figure 19*). When running the remaining passes, use stringer beads and the freehand torch-handling technique. Ensure proper fusion at the toes of the weld bead. Take special care to fill the crater at the termination of the weld.

Follow these steps to practice pipe welds in the 6G position:

Step 1 Tack-weld together the practice coupon, as explained earlier.

Step 2 Clamp or tack-weld the coupon into position with the pipe axis inclined 45° to the horizontal plane.

Step 3 Run the root pass uphill with a stringer bead, as shown in *Figure 16*. Repeat the procedure for the opposite side of the pipe.

Step 4 Brush the root pass to clean the weld.

Step 5 Run the remaining passes uphill at the proper work angles. Overlap the passes and clean the weld after each pass.

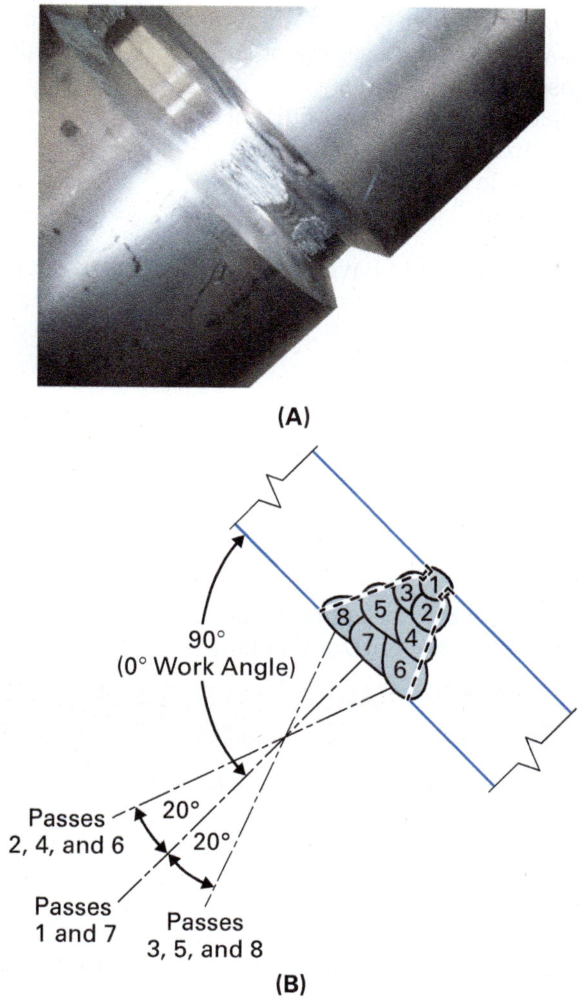

Figure 19 Multipass 6G bead sequences and work angles for the U-groove joint.
Source: Terry Lowe (19A)

1.0.0 Section Review

1. Which of the following statements about welding safety is *true*?
   a. Oxygen is a good and safe choice for blowing dust from clothing.
   b. Wearing ear plugs at any time while welding creates an unacceptable safety issue.
   c. A neat, clean work area reduces hazards and distractions.
   d. GTAW can be done using the lowest level of shading available in the welding hood lens.

2. What is the recommended torch-handling technique used to perform GTAW?
   a. On-the-wire
   b. Freehand
   c. Side-to-side
   d. Walking-the-cup

3. The preferred GTAW joint for aluminum pipe without a backing is the _____.
   a. double U-groove joint
   b. double V-groove joint
   c. open-root V-groove joint
   d. U-groove joint

4. When setting an inverter-type welding machine for use on clean aluminum, a wave balance setting of 80% EN or higher allows the welder to use a(n) _____.
   a. smaller electrode at a higher current
   b. larger electrode at a lower current
   c. blunt electrode
   d. EWP electrode

5. Pipe groove welds should have slight reinforcement, not exceeding _____.
   a. $\frac{1}{32}$" (0.8 mm)
   b. $\frac{1}{16}$" (1.6 mm)
   c. $\frac{3}{32}$" (2.4 mm)
   d. $\frac{1}{8}$" (3.2 mm)

6. When welding in the 5G position, clamp or tack-weld the weld coupon into position with the pipe axis _____.
   a. vertical
   b. horizontal
   c. inclined at 5° to the horizontal plane
   d. inclined at 45° to the horizontal plane

Module 29403 Review Questions

1. The *most* important piece of equipment carried by welders is _____.
   a. fall protection equipment
   b. leather welding gloves
   c. steel-toe boots
   d. a welding hood

2. The torch travel angle that is normally used for GTAW of aluminum is a _____.
   a. push angle of 10° to 15°
   b. push angle of 15° to 20°
   c. drag angle of 10° to 15°
   d. drag angle of 15° to 20°

3. When you are making groove welds on aluminum pipe, the weld puddle size can be increased using a small _____.
   a. uphill motion
   b. downhill motion
   c. circular motion
   d. up and down motion

4. When welding aluminum pipe, the preferred welding table surface is _____.
   a. low-alloy steel
   b. high-carbon steel
   c. aluminum
   d. stainless steel

5. When welding aluminum, the high thermal conductivity of the base metal requires the use of _____.
   a. higher welding currents
   b. lower welding currents
   c. higher welding voltage
   d. lower welding voltage

6. One disadvantage of using helium as a shielding gas, alone or as part of a mixture, is that the _____.
   a. arc may be more difficult to start
   b. melting point of the aluminum rises
   c. aluminum workpiece is blackened
   d. travel speed is reduced

7. The shielding gas to use to increase welding speed and achieve deeper penetration is a mix of _____.
   a. helium and nitrogen
   b. argon and carbon dioxide
   c. helium and carbon dioxide
   d. argon and helium

8. The U-groove weld is preferred for joints without backing because it _____.
   a. allows for a larger weld puddle
   b. prevents complete root penetration
   c. allows for a larger gap in the root opening
   d. reduces heat transfer away from the weld joint

9. Pipe groove weld positions are determined by the _____.
   a. axis of the pipe
   b. length of the pipe
   c. weight of the pipe
   d. thickness of the pipe

10. When welding in the 6G position, clamp or tack-weld the pipe weld coupon into position with the pipe axis inclined _____.
    a. 10° to the horizontal plane
    b. 15° to the horizontal plane
    c. 30° to the horizontal plane
    d. 45° to the horizontal plane

Answers to odd-numbered questions are found in the Review Question Answer Keys at the back of this book.

Answers to Section Review Questions

Answer	Section Reference	Objective
Section 1.0.0		
1. c	1.1.2	1a
2. b	1.2.3	1b
3. d	1.3.2	1c
4. a	1.4.0	1d
5. c	1.5.2	1e
6. b	1.6.2	1f

MODULE 29405

Soldering and Brazing of Copper

Source: © Miller Electric Mfg. LLC

Objectives

Successful completion of this module prepares you to do the following:

1. Describe how to properly prepare and solder copper tubing.
   a. Describe the proper use of the PPE, tools, and materials needed to solder copper tubing.
   b. Describe how to prepare materials for soldering.
   c. Describe how to solder copper tubing.
2. Describe how to properly prepare and braze copper tubing.
   a. Describe the proper use of the PPE, tools, and materials needed to braze copper tubing.
   b. Describe how to prepare the equipment and materials for brazing.
   c. Describe how to braze copper to copper as well as to dissimilar metals.

Performance Tasks

Under supervision, you should be able to do the following:

1. Properly set up and shut down air-acetylene equipment.
2. Properly prepare and solder copper tubing in various planes, using various fittings.
3. Properly set up and shut down oxyfuel equipment.
4. Properly prepare and braze copper tubing in various planes, using various fittings.

Overview

Soldering and brazing are welding processes that use nonferrous filler metal to form the weld joint. Soldering uses filler metal with a relatively low melting point. For joints that need to be mechanically stronger and more pressure-resistant, silver or copper alloy brazing filler metal with a relatively high melting point is used. Oxyacetylene or air-acetylene fuel gases are typically used for brazing. Both soldering and brazing demand careful attention to safety due to the hazards associated with extreme heat, flammable gases, and open flames.

Digital Resources for Welding

NCCER SCAN ME

Scan this code using the camera on your phone or mobile device to view the digital resources related to this craft.

1.0.0 Soldering

Performance Tasks

1. Properly set up and shut down air-acetylene equipment.
2. Properly prepare and solder copper tubing in various planes, using various fittings.

Objective

Describe how to properly prepare and solder copper tubing.

a. Describe the proper use of the PPE, tools, and materials needed to solder copper tubing.
b. Describe how to prepare materials for soldering.
c. Describe how to solder copper tubing.

Soldering: A method of joining metals with a nonferrous (see *nonferrous*) filler metal using heat below 800°F (427°C) and below the melting point of the base metals being joined. Also known as *soft soldering* or *sweating*.

Brazing: A method of joining metals with a nonferrous (see *nonferrous*) filler metal using heat above 800°F (427°C) but below the melting point of the base metals being joined.

Nonferrous: Relating to a group of metals and metal alloys that contain no iron.

Capillary action: The movement of a liquid along the surface of a solid in a spreading action.

Solder: A fusible alloy used to join metals, with a melting point below 800°F (427°C).

Fillet: A rounded internal corner or shoulder of filler metal that often appears at the meeting point of a piece of tubing and a fitting when the joint is soldered or brazed.

The most common thermal methods used for joining copper tubing and fittings are **soldering** and **brazing**. Both methods bond the metals together using a **nonferrous** filler metal that adheres to the surfaces being joined. The filler metal is drawn into and distributed between the closely fitted surfaces by **capillary action**.

The difference between soldering and brazing is the temperature needed to melt the filler metal in order to make the joint. Soldering uses filler metals that melt at temperatures below 800°F (427°C), usually in the 375°F to 500°F (191°C to 260°C) range. The filler metals used in brazing melt at temperatures above 800°F (427°C). The dividing line temperature that defines the two processes has been established by the American Welding Society (AWS).

Soldering, also referred to as *sweating* or *soft soldering*, is a common method of joining copper tubing and fittings for water lines and other low-pressure applications. Soldered joints are generally used in piping systems that carry liquids at temperatures of 250°F (121°C) or below.

Soldering involves using heat and a nonferrous filler metal to join two metal surfaces. A nonferrous filler metal contains no iron and is therefore nonmagnetic. The melting point of the filler metal must be lower than that of the two metals being joined.

Soldered joints depend on capillary action to pull and distribute the melted **solder** into the small gap between the fitting and the tubing. When the joint is filled, the solder will form a tiny bead around the joint called a **fillet**. Capillary action describes the flow of liquid, solder in this case, into a small space between two surfaces. This action is most effective when the space between the tubing and the fitting is between 0.002" and 0.005" (0.05 mm and 0.13 mm). To maintain the proper spacing, it is important to check the joint for correct alignment before soldering. *Figure 1* shows how capillary action works.

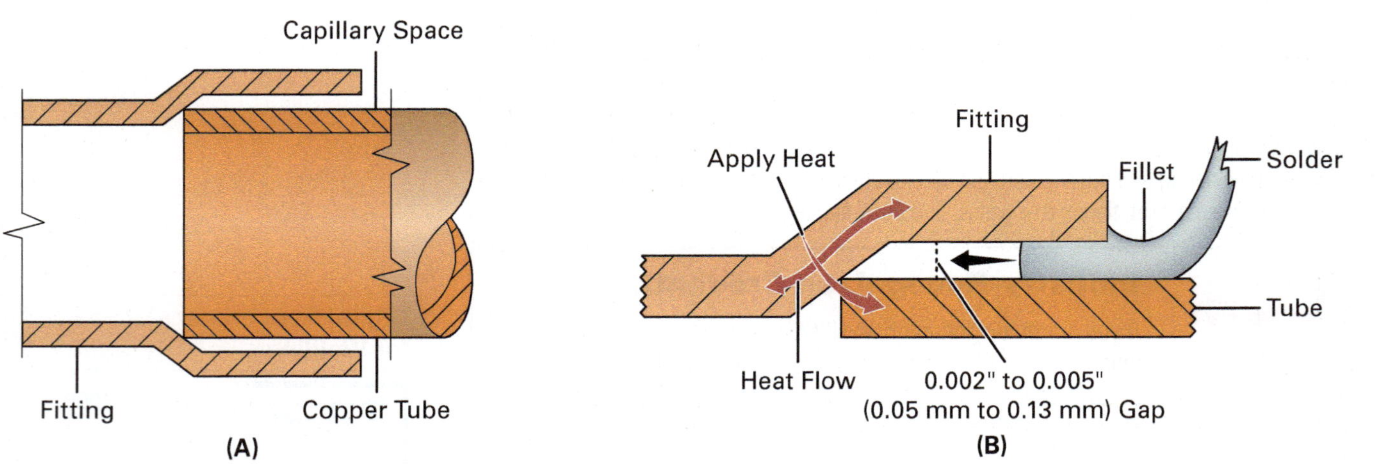

Figure 1 Capillary action.

Did You Know?

Sweating Pipe

To solder copper fittings, the fitting must be heated until the soldering paste starts to melt. The melting paste looks like beads of sweat on the pipe, which led to the term *sweating* to describe the soldering process.

1.1.0 PPE, Tools, and Materials

Soldering does not require a lot of tools or sophisticated equipment. The equipment is relatively simple and easy to operate as long as safe work habits are maintained. As with any skill, the process requires some practice to master.

1.1.1 Soldering Safety

Soldering involves the use of a flammable gas under pressure, an open flame, and possible exposure to molten metal. To avoid injury, workers performing soldering must wear appropriate personal protective equipment (PPE) and observe the following safety procedures:

- Wear clothing made of flame-resistant fabric, such as cotton. Avoid synthetic fabrics such as nylon, which can melt and adhere to the skin.
- Wear a long-sleeve shirt to help prevent burns from sparks or molten metal. Pants should not have cuffs which can catch and hold hot materials.
- Wear eye and face protection. Soldering should be performed using safety glasses as well as a face shield, one of which should be tinted. For soldering, a #2 tint is recommended. Refer to *OSHA Standard 29 CFR 1926.102*.
- Wear flame-resistant gloves and high-top work boots.
- Make sure the gas cylinder is securely turned off when not in use. Gases used in soldering are flammable and explosive if exposed to a spark or flame.
- An appropriate fire extinguisher must be kept close by while soldering in case material near the work ignites. When working close to flammable material such as wood, use a fire blanket or other flame- or heat-blocking material to protect it.

Flux: A chemical substance that prevents oxides from forming on the surface of metals as they are heated for soldering, brazing, or welding.

- Always solder in a well-ventilated area. Fumes from the **flux** can irritate your eyes, nose, throat, and lungs. Wear respiratory equipment if working in an area with poor ventilation.
- Follow specific site safety procedures for the use of an open flame. When working at field sites, acquire and follow all applicable safety guidelines and requirements.

Additional soldering safety information for the construction industry and general industry can be found in *OSHA Standard 29 CFR 1926.350* and *OSHA Standard 29 CFR 1910.253*, respectively.

1.1.2 Soldering Tools and Equipment

The heat used for soldering is provided by a torch that is attached to a cylinder of flammable gas. Propane, MAP-Pro® (propylene), and acetylene are commonly used to make soldered joints in copper tubing. Propylene burns slightly hotter than propane, making it better suited for larger joints. Propane and MAP-Pro® are sold in small cylinders, such as the one shown in *Figure 2*, for handheld use. These cylinders are adequate for small projects, such as occasionally joining fittings to water lines or condensate lines.

When soldering larger, more repetitive projects, the use of acetylene is far more common. Acetylene is generally sold by welding supply companies and trade-related distributors. When portability is important, acetylene can be carried in small, easily transportable cylinders. The smaller MC cylinder shown in *Figure 3* holds 10 ft^3 (0.283 m^3); the larger B cylinder holds 40 ft^3 (1.13 m^3). Larger acetylene cylinders are used when portability is not as important.

Figure 2 Handheld fuel gas cylinder.

Figure 3 B and MC acetylene cylinders.

Acetylene cylinders are packed with a porous material that is partially saturated with **acetone** for safer storage. The packing helps prevent high-pressure gas pockets from forming in the cylinder. Acetone is a colorless, flammable liquid. It is added to the cylinder until about 40% of the porous material becomes saturated. The acetone can hold large quantities of dissolved acetylene under pressure—as much as 25 times its own volume.

Acetone: A colorless organic solvent that is volatile and extremely flammable. In reference to the soldering and welding process, it is used as a carrier for acetylene gas in cylinders.

Because it is a liquid, acetone can be drawn from an acetylene cylinder when it is not upright or when the usage rate of the acetylene gas is excessive. The sudden appearance of flaming acetone at the end of the torch while working can be both alarming and hazardous.

Soldering requires a torch to provide the heat. An air-acetylene torch kit is shown in *Figure 4*. Because acetylene cylinders are large and heavy, a connecting hose and bottle fitting is used. The torch assembly with its fuel cylinder cannot be comfortably handheld.

A self-igniting soldering torch is shown in *Figure 5*. This type of torch is designed to be connected directly to a propylene or propane gas cylinder, such as the one shown in *Figure 2*. Pressing the yellow button ignites the air/gas mixture.

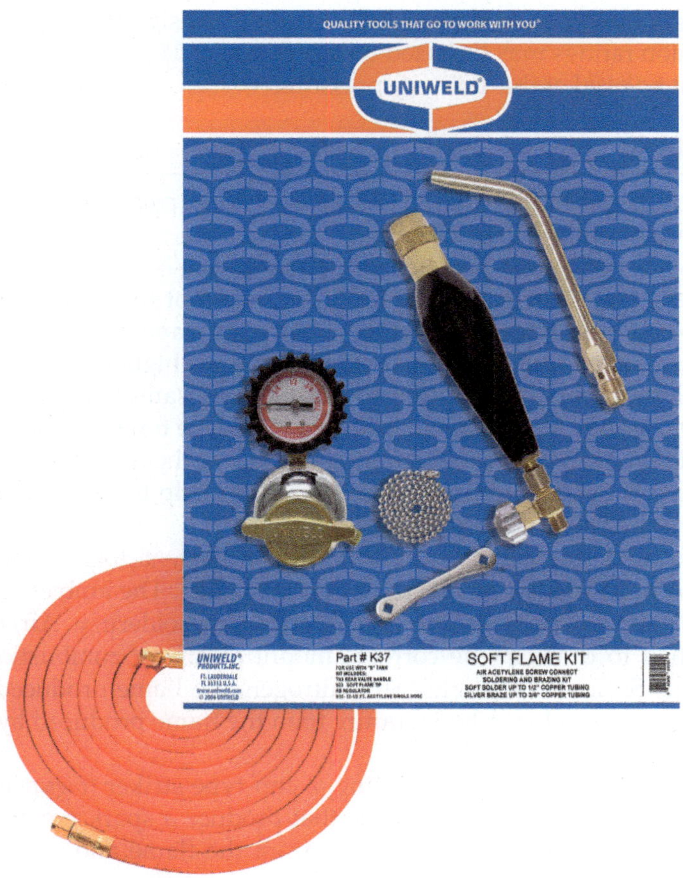

Figure 4 Air-acetylene torch kit.
Source: Uniweld Products, Inc. "Quality Tools That Go To Work With You!"

Figure 5 Self-igniting soldering torch.

Air-acetylene torches can usually accommodate different tips for a smaller or larger flame. The tips are numbered by the manufacturer, with larger numbers typically identifying larger tips. The tip specifications include the actual opening size, the fuel consumption at a given pressure (often rated at 14 psig or 96.5 kPa), and the tubing sizes for which it is recommended. *Table 1* provides an example of the performance data for one manufacturer's series of air-acetylene torch tips.

TABLE 1 Air-Acetylene Torch Tip Performance Data

Tip Model	Tip Size		Gas Flow (SCFH*) at 14 psig	Required Tube Size	
				Up to 180 psi	Over 180 psi
T-2	1/8"	4.8 mm	2.1	1/8" to 1/2"	1/8" to 1/4"
T-3	3/16"	6.4 mm	3.5	1/4" to 1"	1/8" to 1/2"
T-4	1/4"	7.9 mm	5.8	3/4" to 1 1/2"	1/4" to 3/4"
T-5	5/16"	9.5 mm	8.2	1" to 2"	1/2" to 1"
T-7	7/16"	11.1 mm	11.1	1 1/2" to 3"	7/8" to 1 5/8"
T-8	1/2"	12.7 mm	14.4	2" to 3 1/2"	1" to 2"

*Standard cubic feet per hour

Acetylene is a relatively unstable gas. It can become volatile when released from the storage cylinder at pressures exceeding 15 psig (103 kPa). Although fuel consumption ratings may be based on a pressure of 14 psig (96.5 kPa), acetylene should not be used at this delivery pressure.

WARNING!

Never adjust the acetylene regulator to higher than 15 psig (103 kPa). Acetylene becomes volatile above this pressure.

Acetylene has a lower heat content per cubic foot than any other common fuel gas except methane or natural gas. Butane, propane, and propylene all have significantly more heat content per cubic foot. This higher heat content results in much lower fuel consumption to produce the same amount of heat. The major advantage of a properly adjusted air-acetylene torch is that acetylene can achieve higher combustion temperatures than any other gas. Although this high temperature is not required for soldering, it can help to increase productivity when many solder joints must be completed.

Because soldering requires relatively low heat, a torch that mixes the flammable gas directly with air is all that is needed. The ambient air provides enough oxygen to produce the needed heat for soldering. Air/fuel torches are designed to draw in the correct amount of air for combustion from the atmosphere. Since air is roughly 80% nitrogen, the flame produced by an air/fuel torch burns at a lower temperature than the flame produced by an oxyfuel torch.

For torches that are not self-igniting, a cup-type striker (*Figure 6*) should be used to light the torch. Similar styles that produce a spark are also available. Strikers keep the hands a safe distance from the flame and do not contain a flammable substance. Never use a match or pocket lighter to light a torch.

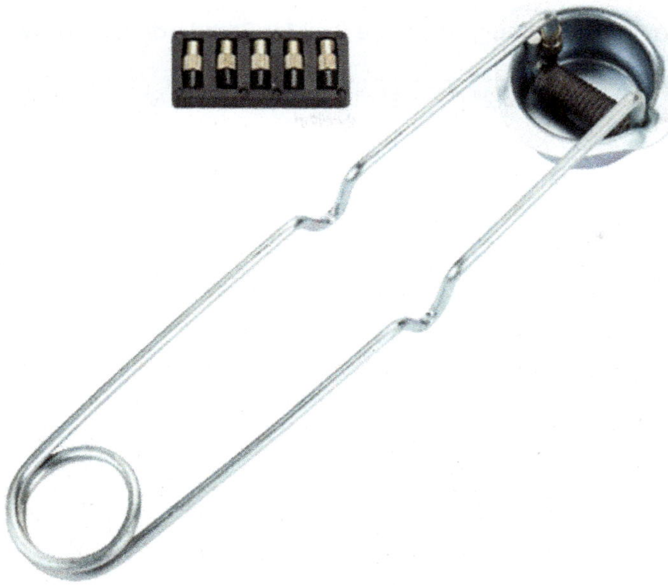

Figure 6 A cup-type striker.
Source: Photo courtesy of The Lincoln Electric Company, Cleveland, OH, U.S.A.

1.1.3 Solders and Soldering Fluxes

Solder is a nonferrous metal or metal **alloy** with a melting point below 800°F (427°C). An alloy is any substance made up of two or more metals. Solder selection is based primarily on the operating temperature and pressure of the piping system. If the system experiences dramatic temperature changes, thermal stress may also be a selection factor.

Alloy: Any substance made up of two or more metals.

It is important to determine whether the system will carry potable water. Solders containing more than 0.2% lead were banned for use on potable water systems by a 1986 amendment to the federal *Safe Drinking Water Act*. The most common solder used on water lines and other low-pressure applications is an alloy made of 95% tin and 5% antimony. *Table 2* provides a comparison of pressure-temperature ratings for four different solder alloys. The 50-50 alloy of tin and lead provides very limited strength and is only suitable for low- or non-pressure applications. It cannot be used on potable water lines.

Solders composed of tin-antimony, tin-copper-silver, or tin-silver are usually recommended for applications requiring greater joint strength. The tin-antimony solder typically melts between 430°F and 480°F (221°C and 249°C) and solidifies rapidly. These solders, in the form of thick yet relatively soft wire, are supplied on spools for easier use (*Figure 7*). The job specifications often identify the type of solder to be used for a specific project.

A flux (*Figure 8*) should be applied to all tube ends and fittings immediately after they are cleaned. Because flux is rather tacky, care must be taken to keep grit and debris off the tube and fittings once flux is applied. Choosing a proper flux is very important as the wrong flux can ruin the soldered joint.

TABLE 2 Pressure-Temperature Ratings of Soldered Joints

Joining Material[d]	Service Temperature (°F)	Fitting Type	Maximum Working Gauge Pressure (psi) for Standard/Nominal Water Tube Sizes[a]				
			1/8" to 1"	1 1/4" to 2"	2 1/2" to 4"	5" to 8"	10" to 12"
Alloy Sn50 50-50 Tin-Lead Solder[e]	100	Pressure[b]	200	175	150	135	100
		DWV[c]	–	95	80	70	–
	150	Pressure[b]	150	125	100	90	70
		DWV[c]	–	70	55	45	–
	200	Pressure[b]	100	90	75	70	50
		DWV[c]	–	50	40	35	–
	250	Pressure[b]	85	75	50	45	40
		DWV[c]	–	–	–	–	–
	Saturated Steam	Pressure	15	15	15	15	15
Alloy Sb5 95-5 Tin-Antimony Solder	100	Pressure[b]	1,090	850	705	660	500
		DWV[c]	–	390	325	330	–
	150	Pressure[b]	625	485	405	375	285
		DWV[c]	–	225	185	190	–
	200	Pressure[b]	505	395	325	305	230
		DWV[c]	–	180	150	155	–
	250	Pressure[b]	270	210	175	165	125
		DWV[c]	–	95	80	80	–
	Saturated Steam	Pressure	15	15	15	15	15

Notes:

For extremely low working temperatures from 0°F to –200°F, it is recommended that a joint material melting at or above 1,100°F be employed[f].

[a]Standard water tube sizes per *ASTM B88*.

[b]Ratings up to 8" in size are those given in *ASME B16.22, Wrought Copper and Copper Alloy Solder-Joint Pressure Fittings*, and *ASME B16.18, Cast Copper Alloy Solder Joint Pressure Fittings*. Ratings for 10" to 12" sizes are those given in *ASME B16.18, Cast Copper Alloy Solder Joint Pressure Fittings.*

[c]Using *ASME B16.29, Wrought Copper and Wrought Copper Alloy Solder-Joint Drainage Fittings – DWV*, and *ASME B16.23, Cast Copper Alloy Solder Joint Drainage Fittings: DWV*.

[d]Alloy designations per *ASTM B32*.

[e]*Safe Drinking Water Act Amendments of 1986* prohibits the use of any solder having a lead content in excess of 0.2% in potable water systems.

[f]These joining materials are defined as brazing alloys by the American Welding Society.

Figure 7 95-5 solder.
Source: Oatey Co.

Figure 8 Two types of flux and a flux brush.
Source: Oatey Co.

Soldering flux performs the following functions:

- It chemically cleans and protects the surfaces of the tubing and fitting from **oxidation**. Oxidation occurs when the oxygen in the air combines with the recently cleaned metal. It produces tarnish or rust in metal and prevents solder from adhering. The oxidizing process is greatly accelerated when the tube and fitting are heated.
- It allows the soldering alloy or filler metal to flow more easily into the joint by promoting **wetting** of the metals.
- It floats out remaining oxides ahead of the molten filler metal.

Oxidation: The process by which the oxygen in the air combines with metal, producing tarnish and rust.

Wetting: A process that reduces the surface tension so that molten (liquid) solder flows evenly throughout the joint.

Fluxes can be classified into three general groups: highly corrosive, less corrosive, and noncorrosive. The soldering process must render the flux inert, meaning the lack of any chemical action. If not rendered inert, the flux gradually destroys the soldered joint over time. The fluxes used for joining copper tubing and copper fittings should be noncorrosive fluxes. Although corrosive types can do a better job of removing oxides and cleaning the copper, they represent a long-term risk to the joint.

One noncorrosive type of flux is composed of water and white rosin dissolved in an organic or benzoic acid base. The best noncorrosive fluxes for joining copper pipe and fittings are compounds of mild concentrations of zinc and ammonium chloride with petroleum bases. Job specifications often identify the type of flux to be used on a specific project. Otherwise, follow the manufacturer's recommendations.

An oxide film begins forming on copper immediately after it has been mechanically cleaned, so it is important to apply flux as quickly as possible after cleaning copper fittings and tubing. Flux should be applied to the clean metal with a brush or swab—never with your fingers. Not only could this cause infection if you have a cut, but flux could possibly be carried to the eyes or mouth as well.

WARNING!

Contact with flux can damage the eyes, causing severe pain, burning, or vision loss. In contact with skin, flux can cause burns, irritation, or necrosis. Gloves should be worn when applying flux. If ingested, flux poisoning can occur.

CAUTION

Brazing flux and soldering flux are not the same product. Do not allow these fluxes to become mixed or interchanged. Combining different types of flux can ruin the soldering or brazing task.

In addition to the health hazards, body contact with cleaned fittings and tubing adds to unwanted contamination of the metal.

Flux must be stirred before each use. If a can of flux is not closed immediately after use or not used for a considerable length of time, the chlorides will separate from the petroleum base.

1.2.0 Preparing Tubing and Fittings for Soldering

To prepare tubing and fittings for soldering, the tubing must be measured, cut, and reamed, and the tubing and fittings must be cleaned. Proper cleaning techniques are required to produce a solid, leakproof joint. Use the following procedure to prepare the tubing and fittings for soldering:

Step 1 If using the face-to-face method, measure the distance between the faces of the two fittings (*Figure 9*).

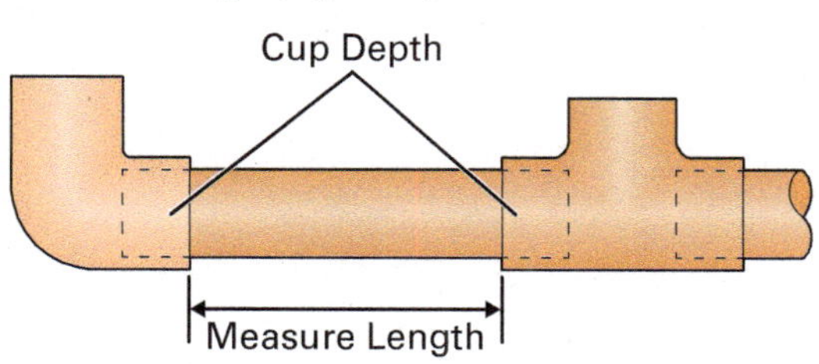

Figure 9 Face-to-face method of measuring pipe length.

Cup depth: The distance that a tube inserts into a fitting, usually determined by a stop inside the fitting.

Step 2 Determine the **cup depth** engagement of each of the fittings. The cup depth engagement is the distance that the tubing penetrates the fitting. This distance can be found by measuring the fitting or by using a manufacturer's makeup chart. *Table 3* shows an example of a manufacturer's makeup chart. Note that metric values are not shown in this example as they are manufactured to different size standards.

TABLE 3 Manufacturer's Makeup Chart

Pipe Size	Cup Depth	Pipe Size	Cup Depth
1⁄4"	5⁄16"	2"	1 1⁄32"
3⁄8"	3⁄8"	2 1⁄2"	1 15⁄32"
1⁄2"	1⁄2"	3"	1 21⁄32"
5⁄8"	5⁄8"	3 1⁄2"	1 29⁄32"
3⁄4"	3⁄4"	4"	2 5⁄32"
1"	29⁄32"	5"	2 21⁄32"
1 1⁄4"	31⁄32"	6"	3 3⁄32"
1 1⁄2"	1 3⁄32"	—	—

Step 3 Add the cup depth engagement of both fittings to the measurement found in *Step 1* to find the length of tubing needed.

Step 4 Cut the copper tube to the correct length using a tubing cutter.

Step 5 Ream the inside of both ends of the copper tube using a reamer.

WARNING!

Do not touch or brush away filings from the tube or fitting with your bare fingers. The ends of cut tubing, especially along the inside edge, are extremely sharp. Always use an appropriate tool. Remember that the natural oils in your fingers will contaminate the freshly cleaned metal.

Step 6 Clean the tubing and fitting to a bright finish using No. 00 steel wool, an abrasive pad, an emery cloth, and/or a tubing brush, as shown in *Figure 10* (A) and *Figure 10* (B). Examine the fitting and tube to ensure they are well-cleaned and no significant flaws exist in the finish (*Figure 10* [C]). Because oxidation may remain in a dent or gouge, flaws in the tubing can usually be seen more easily after cleaning. The tube should be cleaned back from the end slightly farther than the cup depth of the fitting.

Step 7 Stir the flux before applying it. Using a brush or swab, apply a thin film of flux to the copper tubing and to the inside of the copper fitting socket immediately after cleaning the parts. See *Figure 10* (D) and *Figure 10* (E).

Step 8 Insert the tube into the socket, while turning the tube, until the tube touches the inside shoulder of the fitting.

Step 9 Wipe away any excess flux from the joint, as shown in *Figure 10* (F).

Step 10 Check the tube and fitting for proper alignment before soldering.

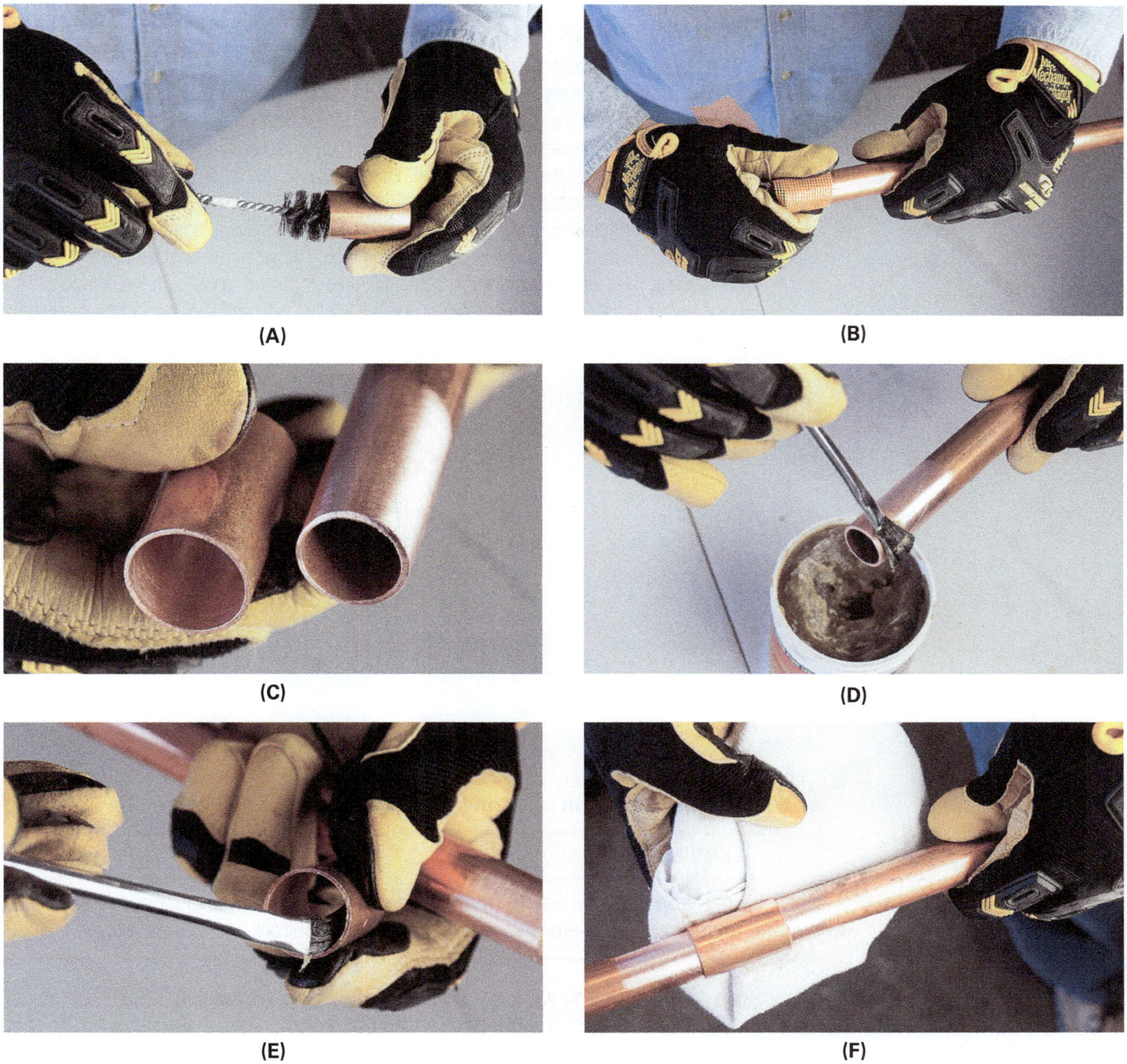

Figure 10 Preparing copper for soldering.

1.3.0 Soldering Joints

With the tubing and fitting properly prepared, the soldering process can begin. Joints that have been prepared should be soldered within a few hours and not left to sit for any length of time. It is best to have the soldering equipment ready so that the soldering can be done as quickly as possible.

Ensure the tubing and fittings to be soldered are well supported. Once the heating process begins and the copper begins to expand, joints tend to loosen and become misaligned. Consider that excess heat can damage certain components, such as valves with seals and other heat-sensitive materials. If a heat-sensitive component cannot be temporarily removed, then it should be protected by being wrapped with a very wet cloth. Heat-absorbing pastes can also be used to protect components.

Heat Sink Products

Overheating when soldering air conditioning and refrigeration piping components is a common issue. Many components that must be soldered or brazed into the piping system, such as reversing valves and thermostatic expansion valves, are also sensitive to the extreme heat applied during the process. Some components can be disassembled before the installation, leaving only the metal body exposed to the heat. However, this is not always possible.

A variety of products are available to help absorb the heat that would otherwise be transferred to the component. Heat sinks are materials that have a high capacity for absorbing heat. Apply the paste or gel in a ring around the tubing between the heat source and the component. If the component itself is being soldered or brazed, apply it on the component body away from the joint. The product should be applied with good contact, without leaving voids that would allow the heat to pass through.

To safely perform soldering on copper tube and fittings, follow these steps:

WARNING!

Fumes from the flux can irritate your eyes, nose, throat, and lungs. Always solder in a well-ventilated area. Wear a respirator if required.

Step 1 Prepare the work area and protect any components in the piping system that could be damaged by the soldering process. Remove any loose, flammable materials from the vicinity and ensure there is adequate room for physical movement around the joint. Place an appropriate fire extinguisher within easy reach and wear necessary PPE.

WARNING!

Whenever soldering or brazing with an open flame, always keep a fire extinguisher of the correct type nearby. Fires can start accidentally if loose materials or wood is ignited by the torch. Shield any combustible materials near the work area with a fire blanket placed a safe distance away.

Step 2 Obtain either an acetylene cylinder and the related equipment (shown in *Figure 11*), or a propane bottle and torch assembly, as well as the required solder. The general instructions provided here are for an air-acetylene torch, but the manufacturer's instructions should always take precedence. As with all equipment used with fuel gases, follow the manufacturer's instructions carefully.

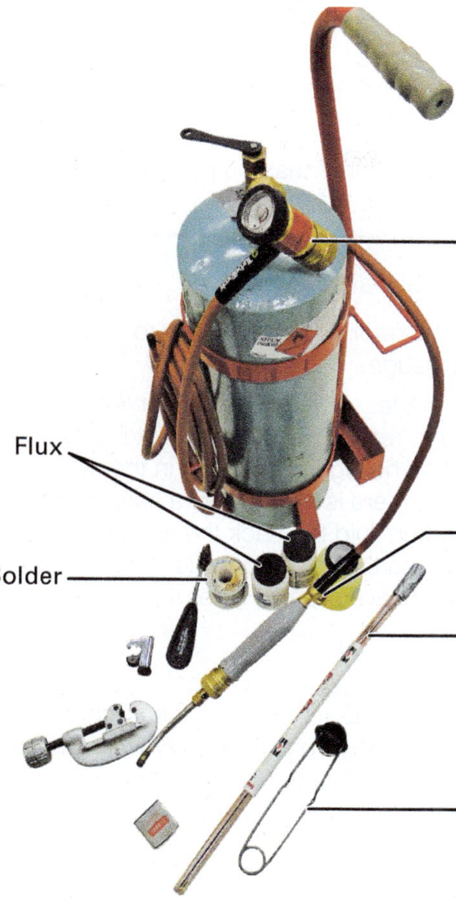

Figure 11 Acetylene B cylinder and related equipment.

WARNING!

Acetylene is a relatively unstable gas. It can become volatile when released from the storage cylinder at pressures exceeding 15 psig (103 kPa). Acetylene regulators should never be set beyond this pressure. To minimize the risk of liquid solvent being drawn into the flow of gas, acetylene should not be intermittently drawn from a cylinder or vessel at a rate greater than 10% of the cylinder's contents per hour. If the flow of acetylene from the cylinder is continuous, the rate of use should not be greater than 7% of the cylinder's contents per hour.

Always keep the cylinder in an upright position and do not exceed the recommended usage rate. Open the cylinder valve only $\frac{1}{2}$ to 1 turn and leave the valve wrench in place on the cylinder at all times. Failure to follow the proper safety procedures can cause a fire, explosion, and/or serious personal injury.

Step 3 Purge (clear out) the acetylene hose as follows:

- Crack the acetylene cylinder valve open slowly until a small amount of pressure registers on the acetylene high-pressure gauge. Then open it about $\frac{1}{2}$ turn.
- Turn the acetylene regulator adjusting screw clockwise until a small amount of pressure shows on the acetylene working-pressure gauge. Allow a small volume of acetylene through to purge the hose, clearing it of any loose debris.
- Turn the acetylene regulator adjusting screw counterclockwise until it is loose. This will shut off the regulator output.

Step 4 Install the brazing torch handle on the ends of the hoses and close the valves on the torch.

WARNING!

Never adjust the acetylene regulator to higher than 15 psig (103 kPa). Acetylene becomes volatile above this pressure.

Step 5 Check the equipment for leaks as follows:

- Adjust the acetylene regulator adjusting screw to 10 psig (70 kPa) on the working-pressure gauge.
- Close the acetylene cylinder valve and check for leaks. If the working-pressure gauge remains at 10 psig (70 kPa) for several minutes, there should be no significant leaks in the system. If the pressure reading drops, there is a leak. Use a soap solution or commercial leak detection fluid to check the acetylene connections for leaks. Never use hoses that are in questionable condition, dried or cracked, or appear to have a leak. Hoses in poor or bad condition should be destroyed and replaced.

Step 6 Light the torch according to the type of equipment you are using and the manufacturer's instructions. Before lighting the torch, don gloves and goggles. Point the torch away from your body when lighting it. Use only an approved lighter (refer to *Figure 6*) to light the torch, unless it is a self-igniting model. Light the torch as follows:

- Open the torch handle valve and then set the acetylene regulator to the required pressure while the gas is flowing. Under no circumstances should the outlet pressure exceed 15 psig (103 kPa).
- Using an appropriate device for ignition, light the torch.

WARNING!

Do not use a match, pocket cigarette lighter, or smoking materials to light the torch. This could result in severe burns or cause the lighter to explode.

- Note that the torch handle valve is not generally used to control the size of the flame. The torch handle valve should be opened fully. The size of the flame is controlled by the torch size and the acetylene pressure regulator setting.

Step 7 Heat the tubing first, with the flame perpendicular (at a right angle) to the tube. Then move the flame onto the fitting cup (see *Figure 12*). Make sure not to overheat the joint. The size of the tubing will determine how much time this step requires.

Step 8 Momentarily move the flame back to the tubing, and then to the back shoulder of the cup. As the flame is moved back to the cup, also back the flame away slightly to avoid overheating.

Step 9 Touch the end of the solder to the area between the fitting and the tube. The solder will be drawn into the joint by capillary action. The solder can be fed upward or downward into the joint. If the solder does not melt quickly on contact with the joint, move the solder away and heat the joint further. Do not melt the solder with the flame; the solder must be melted by the heated copper.

Cup-Type Striker

When using a cup-type striker to ignite a torch, hold the cup of the striker slightly below and to the side of the tip, parallel with the fuel gas stream. This helps prevent the ignited gas from deflecting back toward you from the cup. It also reduces the amount of soot that accumulates in the cup. The flint in most cup-type strikers can be replaced when it is worn out.

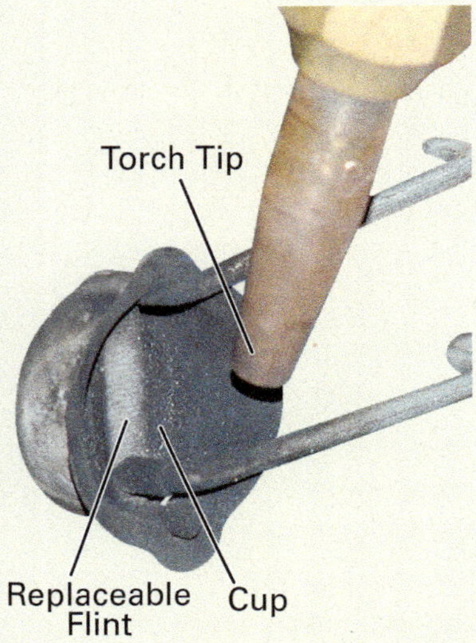

CAUTION

The inner cone of the flame should barely touch the metal being heated. Do not direct the flame into the socket because this will burn the flux. Be sure to keep the flame moving instead of holding the flame in one place.

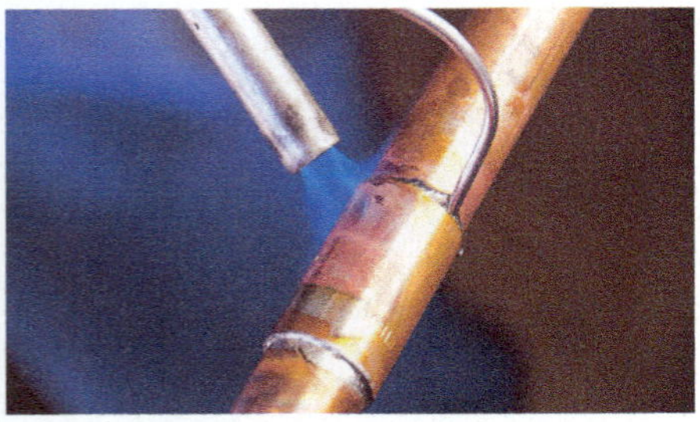

Figure 12 Applying heat to the fitting.

Step 10 Continue to feed the solder into the joint until a ring of solder appears around the joint, indicating that the joint is filled. On nominal $^3/_4$" outside diameter tubing (20 mm metric tubing) and smaller, the solder can generally be fed into the joint from one point. On larger tubing, the solder should be fed quickly around the entire circumference of the joint.

Step 11 Allow the joint to cool until the solder solidifies. While the joint is still warm but not hot, wipe it with a clean damp cloth to remove any excess flux (*Figure 13*). Avoid shocking the joint with water while the joint is very hot as this may cause unnecessary stress and eventual failure.

As a general rule, the amount of solder needed is equal to the diameter of the tubing for sizes 1" (25 mm) and below. For example, with standard $^3/_4$" tubing (20 mm metric tubing), a $^3/_4$" (19 mm) length of solder wire should properly fill the joint. As the size increases above 1" (25 mm metric tubing), more solder will be necessary. For example, a 2" (50 mm metric tubing) joint with an average joint clearance of 0.005" (0.13 mm) will require about $3^1/_2$" (90 mm) of solder. In all cases, the joint clearance significantly affects the amount of solder that is needed.

If the joint is being soldered with the horizontal tubing, the top of the joint will likely heat up on its own and will not need preheating. Concentrate on the bottom and sides to avoid burning the flux at the top. Start feeding the solder near the bottom of the joint and work up toward the top. Then bring the solder back to the starting point and work up the other side toward the top. Remember that, although gravity does influence the solder, capillary action pulls the solder into the joint, rather than gravity.

Soldering is a skill that requires practice. As you watch how the solder and flux behave, you will learn when to move and where to position the torch flame. With practice, you will soon be able to produce well-made solder joints.

Figure 13 Wiping away excess flux.

Copper Fittings

Two types of fittings are available for copper tubing. The first is a wrought copper fitting, which is made from copper tubing that is shaped into different types of fittings. Wrought copper fittings are generally lightweight, smooth on the outside, and have thin walls. The second type is a cast fitting. This type of fitting is made using a mold. The first cast fittings had holes in the sockets to put solder in. Today, the heated copper is poured into the mold and allowed to cool. Cast copper fittings have a rougher exterior and come in a wide variety of shapes. They are typically heavier than wrought copper fittings. Wrought copper fittings are the type used in HVACR work.

1.0.0 Section Review

1. Clothing worn when performing soldering or brazing should be made from flame-resistant fabrics such as _____.
   a. nylon
   b. cotton
   c. polyester
   d. silk

2. The *proper* application of soldering flux involves _____.
   a. using brazing flux if soldering flux is not available
   b. applying the flux only to the tube end using a brush or swab
   c. applying the flux to the tube end and fitting socket using a brush or swab
   d. applying the flux using the fingers or a rag

3. When soldering a joint, _____.
   a. only the tubing is heated
   b. only the fitting is heated
   c. heat is applied directly to the tip of the solder
   d. heat is first applied to the tubing and then to the fitting

2.0.0 Brazing Copper Tubing and Fittings

Performance Tasks

3. Properly set up and shut down oxyfuel equipment.
4. Properly prepare and braze copper tubing in various planes, using various fittings.

Objective

Describe how to properly prepare and braze copper tubing.

a. Describe the proper use of the PPE, tools, and materials needed to braze copper tubing.
b. Describe how to prepare the equipment and materials for brazing.
c. Describe how to braze copper to copper as well as to dissimilar metals.

Brazing, like soldering, uses nonferrous filler metals to join base metals that have a melting point above that of the filler metals. Brazing uses filler metals that melt at temperatures above 800°F (427°C). Brazed tubing and fittings are used in the following applications:

- Low-pressure steam lines
- High-pressure refrigeration lines
- Medical gas lines
- Compressed air lines
- Vacuum lines
- Fuel lines
- Other chemical lines that need extra corrosion resistance in the piping joints

Brazing, sometimes incorrectly referred to as *hard soldering*, produces mechanically strong, pressure-resistant joints. The strength of a brazed joint results from the filler metal's ability to adhere to the base metal. However, reliable adhesion can occur only if the base metals are properly cleaned, the proper flux and filler metal are selected, and the clearance gap between the outside of the tubing and the inside of the fitting is appropriate. Ideally, the clearance is approximately 0.003" to 0.005" (0.08 mm to 0.13 mm).

2.1.0 PPE, Tools, and Materials

To properly braze copper tubing and fittings, it is necessary to be familiar with each of the following aspects:

- Required PPE and safety practices
- Brazing equipment and accessories
- Filler metals and fluxes
- Tube and fitting preparations
- Setting up and lighting a torch
- Nitrogen purging equipment and use
- Brazing techniques

2.1.1 Brazing Safety

Brazing is often done with a combination of oxygen and acetylene. Although the hazards are similar to those encountered in soldering, the hazards are compounded through the use of two gases, each of which can be hazardous if handled improperly. In addition, when fuel gases are directly combined with oxygen instead of air, the temperature of the flame rises significantly. The following safety guidelines apply to brazing:

- Wear clothing made of flame-resistant fabric, such as cotton. Avoid synthetic fabrics such as nylon, which can melt and adhere to the skin.

- Wear a long-sleeve shirt to avoid burns from sparks or molten metal. Pants should not have cuffs.
- Wear eye and face protection. Brazing should be performed using safety glasses as well as a face shield, one of which should be tinted. For brazing, a #3 or #4 tint is recommended. Refer to *OSHA Standard 29 CFR 1926.102*.
- Wear flame-resistant gloves and high-top work boots.
- Make sure gas bottles are securely turned off when not in use. Gases used in brazing are flammable and explosive if exposed to a spark or flame.
- An appropriate fire extinguisher must be kept close by while brazing in case material near the work ignites. When working close to flammable material such as wood, use a fire blanket or other flame- or heat-blocking material to protect it.
- Brazing fumes can irritate your eyes, nose, throat, and lungs. Always braze in a well-ventilated area. Wear respiratory equipment, if necessary.
- Acetylene cylinders must be stored, transported, and used in the upright position. Otherwise, liquid acetone could be pulled out of the cylinder. An acetylene cylinder that has been laid down should be placed in the upright position for at least 30 minutes before using. This allows the solvent to settle back into the porous material and prevents its discharge while the torch is in use. When not in use, cylinders must be stored in the upright position with a safety cap installed, as shown in *Figure 14*. Some small acetylene cylinders, such as the MC cylinder, cannot be capped. When in transit, the regulator must be removed from cylinders and the safety cap securely installed. In many states, it is against the law to transport cylinders with the regulators in place.
- Cylinders must be secured to a cart or structure using a safety chain or approved strap.
- Oxygen and fuel gas cylinders must be stored at least 20' (6.1 m) apart or separated by a wall at least 5' (1.5 m) high with a 30-minute minimum fire rating.
- Oxygen can ignite even when there is no obvious flame or spark to set it off, especially when oxygen comes in contact with oil or grease. Never handle oxygen cylinders with oily hands or greasy gloves. Keep grease away from the cylinders and do not use oil or grease on cylinder attachments or valves. Never use an oxygen regulator for any other gas or try to use a regulator with oxygen that has been used for other service.
- A pressure-reducing regulator set for not more than 15 psig (103 kPa) must be used with acetylene. Acetylene becomes unstable and volatile above this pressure.
- Do not allow anyone to stand in front of the oxygen cylinder valve when opening it. The oxygen is under high pressure (roughly 2,000 psig or 140,000 kPa) and could cause severe injury when the pressure is released.
- Acetylene cylinder valves should only be opened $\frac{1}{2}$ to 1 turn, and the wrench should never be removed from the valve during use. The valve on oxygen cylinders, however, should always be opened fully.

Figure 14 Acetylene cylinder with a valve safety cap.

Transporting Fuel Gas Cylinders

Although OSHA has specific standards for the handling of fuel gas cylinders in *OSHA Standard 29 CFR 1910.253*, the Department of Transportation has primary jurisdiction over the cylinders themselves and their transportation requirements. These requirements are contained in *49 CFR, Parts 171 through 177*. Acetylene must be transported in well-ventilated vehicles and properly secured in an upright position at all times. One interesting provision under the standards is that when drivers are transporting hazardous fuel gases of a quantity that requires placards to be placed on the vehicle, they are prohibited from both texting and cell phone use while driving.

2.1.2 Brazing Equipment

Brazing is typically done using either an acetylene B cylinder with an air-acetylene torch, or an oxyacetylene setup (*Figure 15*). A complete B cylinder setup was previously shown in *Figure 11*. The air-acetylene setup is generally used on smaller line sizes, since it produces less heat than the oxyacetylene equipment. For lines greater than 1" (25.4 mm) in diameter, the oxyacetylene method is usually selected. It provides both heat and better control over the movement of the filler metal. Oxyacetylene equipment mixes acetylene, as a fuel gas, with pure oxygen to produce a hotter flame.

The equipment shown in *Figure 15* is mounted on a hand truck and is very heavy. This type of equipment is typically used in a shop or on a jobsite when extensive brazing must be accomplished in a limited area. For small installations and service-related brazing, portable equipment that can be hand-carried by one person is generally used (*Figure 16*).

Oxygen and acetylene are compressed and shipped under medium-to-high pressures in cylinders. Oxygen is supplied in cylinders at pressures of about 2,000 psig (140,000 kPa). Acetylene cylinders are pressurized to about 250 psig (1,725 kPa). These cylinders must not be moved unless the protective caps are in place. Dropping a cylinder without the cap installed may break the valve off the cylinder, allowing the pressure to escape and propelling the cylinder like a rocket.

Regulators (*Figure 17*) must be installed on oxygen and acetylene cylinders before they can be used. The regulator reduces the high pressure in the cylinder to the desired pressure for use. The correct regulator must be selected for the gas being used. The connecting fittings are typically designed to prevent a regulator from being installed on the wrong gas cylinder. Different cylinder sizes of the same gas may also have different regulator connections. Regulators must be handled carefully. Avoid dropping or otherwise shocking them.

WARNING!

A regulator should always be installed on a cylinder before the valve is opened. The high pressure inside the cylinder, if released suddenly, can cause the cylinder to become a dangerous projectile. Although it is a generally accepted practice to slightly open (crack) the valve of a cylinder to blow out any debris before attaching a regulator, it must be done with great care to ensure the cylinder does not tip over in reaction to the release.

Many regulators have two stages of pressure regulation. The first stage reduces the cylinder pressure to an intermediate level and has a fixed pressure set point. The second stage of the regulator reduces the intermediate pressure to the outlet pressure desired by the user and is adjustable from 0 psig (0 kPa) to the maximum outlet pressure specified by the manufacturer. The final adjustment of gas flow volume at the nozzle is set using the valves on the torch handle.

A regulator usually has two pressure gauges. One gauge shows the pressure in the cylinder, while the other shows the outlet pressure set point. The outlet gauge is monitored as the user sets the desired outlet pressure. The regulator is adjusted by turning the adjusting screw clockwise (turning it in).

Some simpler or less expensive regulators, such as those used on small acetylene cylinders, provide only one stage of regulation. A single-stage regulator often needs adjustment more often, as the outlet pressure tends to fall as the pressure in the bottle changes. Two-stage regulators generally prevent this from happening. Small single-stage acetylene regulators often do not show the pressure in the cylinder but instead have a less accurate gauge showing the volume that remains.

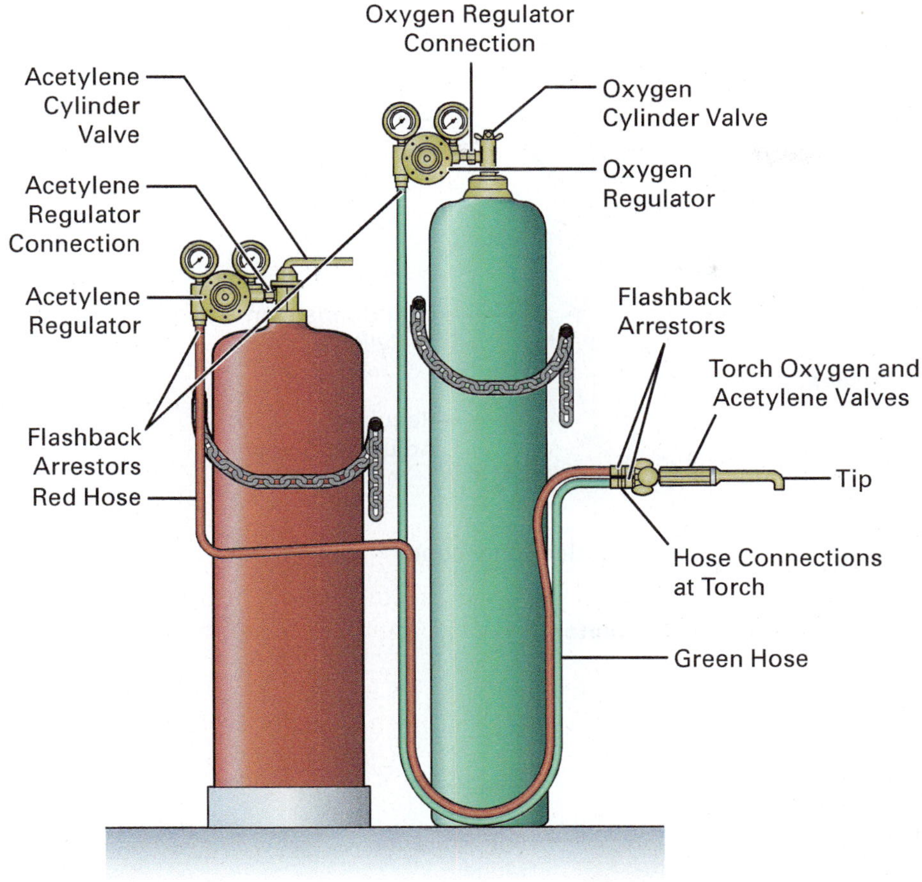

Figure 15 Oxyacetylene brazing setup.

Figure 16 Portable oxyacetylene equipment.

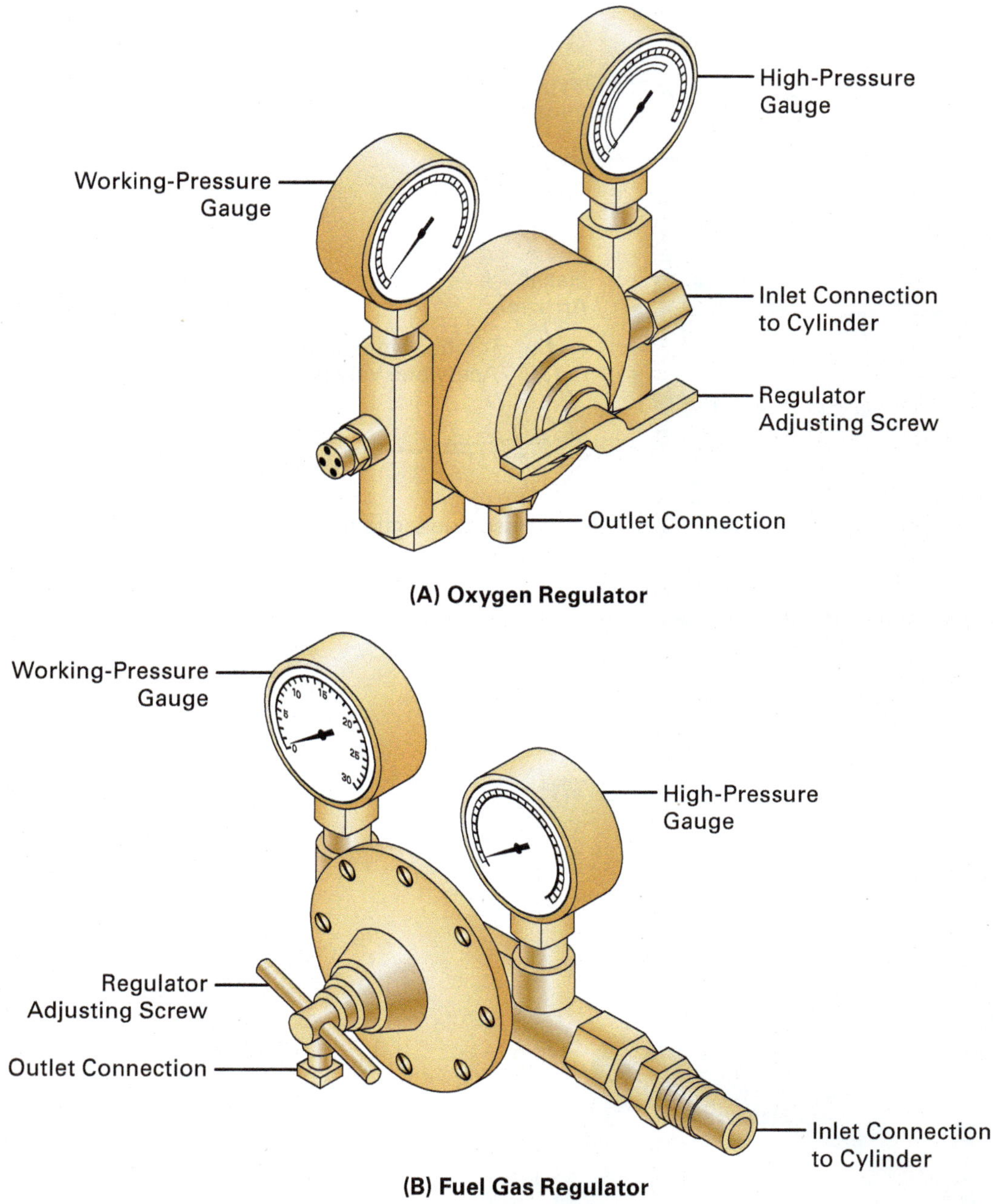

Figure 17 Oxygen and fuel gas regulators.

Hoses designed specifically for brazing and welding work must be used to connect the regulator outlets to the torch handles. The hoses are color coded for visual identification. In the United States, the oxygen hose is green and the fuel gas hose is red. Other countries may have different standards and associated color assignments. The color coding of brazing and welding hoses is an important feature, as the hose fittings are designed so that they can only be connected to the proper gas regulator and torch handle port. Oxygen hoses have a right-handed thread, while fuel gas hoses use a left-handed thread. The female fittings with left-handed threads have a groove cut into the nut for further visual identification. New hoses should be purged with high-pressure air before use, as they often contain a white powder.

Figure 18 shows a universal torch wrench designed for regulators, hose connections, check valves, **flashback arrestors**, torches, and torch tips. The fittings for torch equipment are usually brass or bronze, and certain components are often fitted with soft, flexible, O-ring seals. The seal surfaces of the fittings or O-rings can be easily damaged by overtightening with standard mechanic's wrenches. The length of a torch wrench is limited to reduce the risk of damage to fittings from application of excessive torque. In some cases, manufacturers specify only hand tightening for certain fitting connections of a torch set (tips or cutting/welding attachments). Always follow the manufacturer's specific instructions when connecting the components of a torch set. The torch wrench can be tethered to the torch outfit so that it is always available when needed and not easily misplaced.

Flashback arrestors: A type of valve that prevents a flame from traveling back from the torch tip and into the hoses.

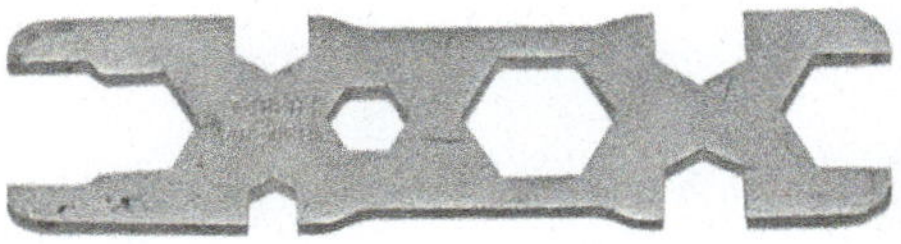

Figure 18 Torch wrench.

Oxyacetylene brazing equipment offers a wide array of torch handles and tips to choose from. The style and size of the tip is determined by the size of the tubing and, to a lesser degree, the type of joint to be made. Brazing torch handles are offered in heavy-, medium-, and light-duty models. For most field brazing applications, the greater maneuverability of medium- and light-duty handles are generally preferred over heavy-duty models. The handle typically provides a mixing chamber for the two fuel gases before they reach the nozzle.

Figure 19 shows examples of light- and medium-duty brazing torch handles. Note that the fuel gas adjusting valves are above the user's hand on a light-duty handle and below the hand on a medium-duty handle. Both of these traits affect the balance of the torch. Having the adjusting valves above the hand can be more convenient, but they can also get in the way in tight spots.

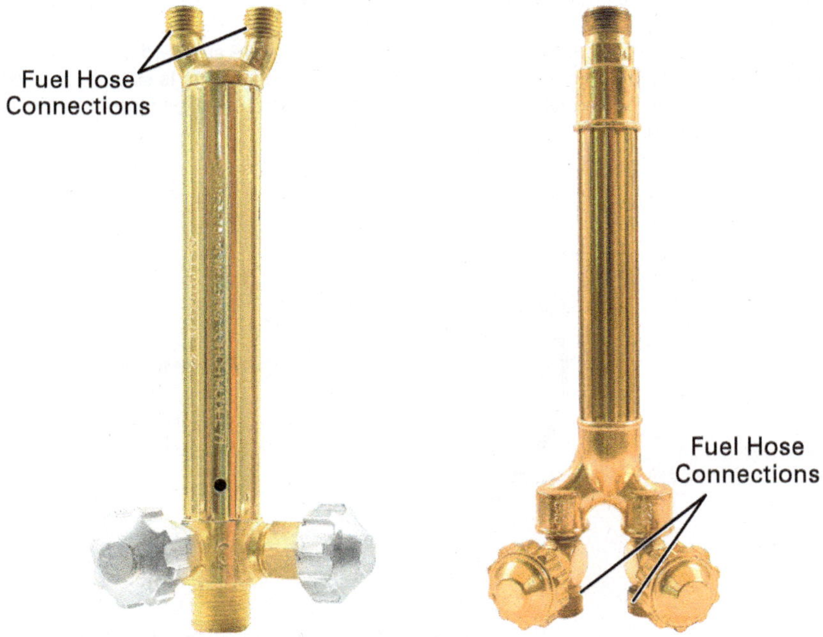

Figure 19 Light- and medium-duty brazing torch handles.
Source: Uniweld Products, Inc. "Quality Tools That Go To Work With You!"

Many torch handles today have built-in flashback arrestors. Flashback arrestors are one-way valves that prevent a pressurized flame from traveling back up the tip and into the hoses and regulators. A flashback can sometimes happen if either the oxygen or the fuel gas flow rate is inadequate, or if both the oxygen and fuel gas are turned on and then ignited at the brazing tip. If a torch handle is not already equipped with flashback arrestors, they should be installed where the hoses connect to the torch handle. Although regulators can be equipped with them at their outlet, torch-mounted models prevent a flashback from entering the hoses. To be effective, flashback arrestors must be installed with the flow arrow pointing toward the torch handle (*Figure 20*).

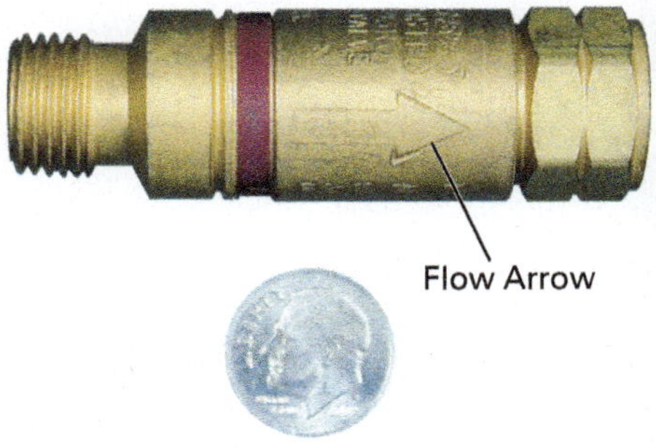

Figure 20 Flashback arrestor.

The brazing tip must be designed to properly fit the handle. Most brazing tips from a given manufacturer are designed to fit only their own handles. Further, tips do not fit every handle that a manufacturer offers. When selecting and purchasing tips, always be sure that they are compatible with the torch handle in use.

Different brazing tips are shown in *Figure 21*. Brazing nozzles and tips are sized by a number corresponding to the size of the nozzle opening. Larger numbers indicate larger openings. *Table 4* provides a guide for the selection of tips based on the size of the tubing to be brazed. Much smaller tips are available but not typically used for commercial trade work.

TABLE 4 Tip Sizes Used for Common Tubing Sizes

Tip Size (No.)	Rod Size		Tubing Diameter	
	in	mm*	in	mm**
4	3/32	2.4	1/4–3/8	10–12
5	1/8	3.2	1/2–3/4	17–22
6	3/16	4.8	1–1 1/4	28–35
7	1/4	6.4	1 1/2–2	42–54
8	5/16	7.9	2–2 1/2	54–64
9	3/8	9.5	3–3 1/2	76–89
10	7/16	11.1	4–6	108–159

Notes:

*Metric rod sizes shown are roughly equivalent to the US standard sizes shown; metric rod may be available in other sizes.

**Metric copper tubing diameters are significantly different than US standard sizes. Sizes shown in inches are based on tubing inside diameter (ID); metric copper tubing is sized by outside diameter (OD). Therefore, the metric OD values shown represent the sizes closest to the US standard ID sizes shown.

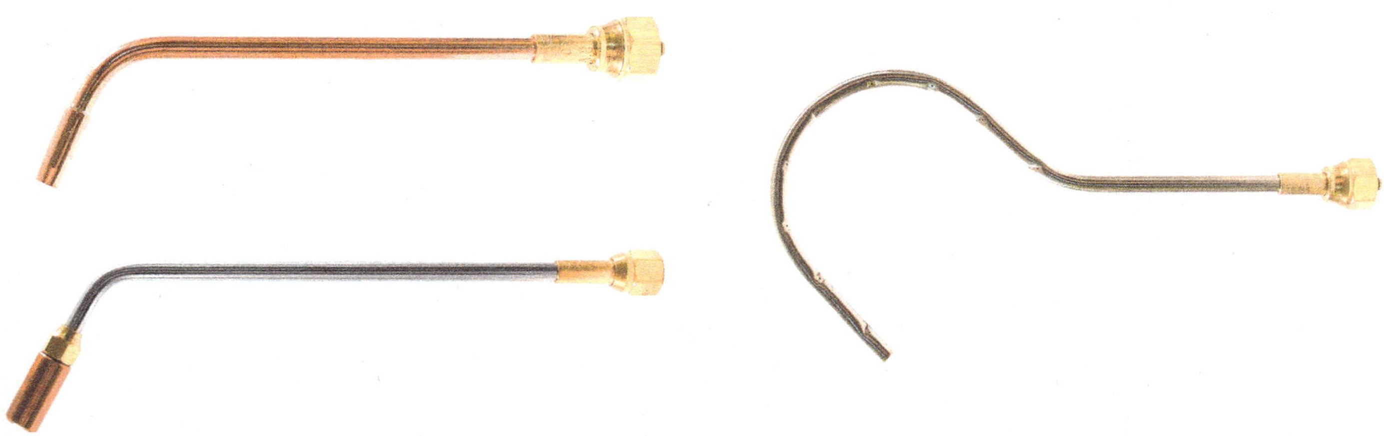

(A) Multiflame (Rosebud) Torch Tip

(B) Multiflame Wraparound Torch Tip

Figure 21 Brazing tips.
Source: Uniweld Products, Inc. "Quality Tools That Go To Work With You!"

The multiflame, or rosebud, torch tip in *Figure 21* is used for larger tubing and to heat larger surfaces. Instead of a single opening in the nozzle, several smaller openings are provided. The rosebud is an excellent choice for larger tubing and fittings to broaden the area of the flame.

The multiflame wraparound tip, also shown in *Figure 21*, consumes a significant amount of fuel and oxygen, but the design is excellent for brazing smaller tubing. The wraparound design ensures that the joint is equally hot all around and brazed joints can be completed very quickly.

Brazing tips occasionally get fouled with solids from the combustion process. This is especially true when the torch is not set up to burn cleanly. The relatively soft brass can also be damaged at the tip, causing the flame to be erratic. Torch tip cleaners (*Figure 22*) are used to clean the tip. They are like small, round files that clear the material out of the way. Each tip cleaner fits a specific nozzle size. As shown, the tip cleaners typically come as a set in folding packages.

Tip drills are used for major cleaning and for holes that are plugged. These are tiny drill bits that are sized to match the diameters of the tip openings. A drill bit is mounted in a small handle for manual use.

Figure 22 Torch tip cleaners.
Source: Uniweld Products, Inc. "Quality Tools That Go To Work With You!"

Tiny Tips

Oxyacetylene torch tips are available with extremely tiny openings for jewelry work and other precision applications. Some nozzles are so small that the level of precision needed cannot be achieved by drilling the nozzle opening in metal. Instead, a hole as small as 0.003" is drilled into a precious stone, such as a sapphire, and the stone is then inserted into the tip.

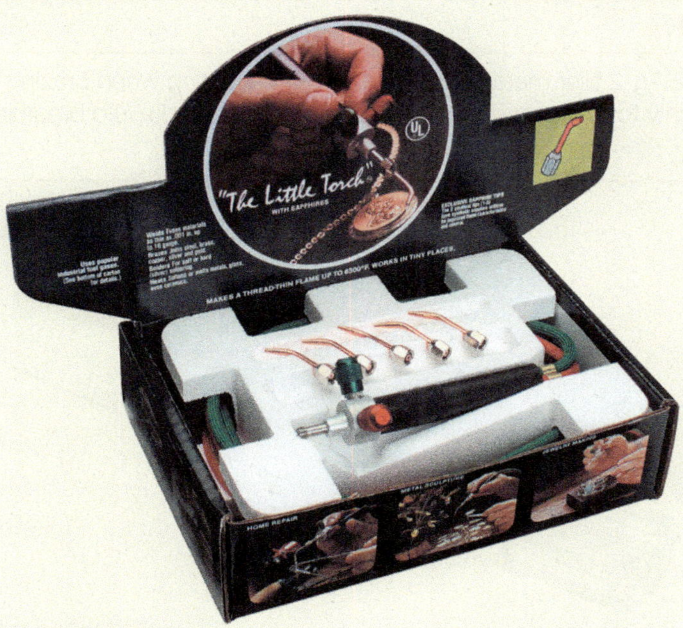

Source: © Miller Electric Mfg. LLC

2.1.3 Filler Metals and Fluxes

Filler metals used to join copper tubing are comprised of two groups: alloys that contain 8% to 60% silver (the BAg series) and copper alloys that contain phosphorus (the BCuP series). *Table 5* lists brazing filler materials and principal elements according to the American Welding Society (AWS) classification. *Figure 23* shows a package of BCuP-5 brazing filler rods that contain 15% silver, 80% copper, and 5% phosphorous. They have been an industry standard for many years, although other alloys have dramatically increased in popularity as silver prices have become less affordable. Unlike soldering filler metals, brazing filler metals are typically in the form of a stick.

The two major groups of filler metals differ in their melting, fluxing, and flowing characteristics. These characteristics should be considered when selecting a filler metal. When joining copper tubing, any of the filler metals in *Table 5* can be used. However, the filler metals used most often for close tolerances and reasonable economy are BCuP-3 and BCuP-4. BCuP-5 is better where close tolerances cannot be held in the joint, but it is generally a more expensive product.

Flux is used in brazing as it is for soldering, but it is not always needed. Alloys that contain phosphorus, such as BCuP-3 through BCuP-5, are self-fluxing when used in clean copper-to-copper applications. When they are used with brass, bronze, or ferrous metals, a flux is required.

TABLE 5 Brazing Filler Metals

Percentage of Principal Element						
AWS Classification	**Silver**	**Phosphorus**	**Zinc**	**Cadmium**	**Tin**	**Copper Classification**
BCuP-2	—	7–7.5	—	—	—	Balance
BCuP-3	4.75–5.25	5.75–6.25	—	—	—	Balance
BCuP-4	5.75–6.25	7–7.5	—	—	—	Balance
BCuP-5	14.5–15.5	4.75–5.25	—	—	—	Balance
BAg-1	44–46	—	14–18	23–25	—	14–16
BAg-2	34–36	—	19–23	17–19	—	25–27
BAg-5	44–46	—	23–27	—	—	29–31
BAg-7	55–57	—	15–19	—	4.5–5.5	21–23

WARNING!

BAg-1 and BAg-2 filler metals contain cadmium. Heating when brazing can produce highly toxic fumes. Use adequate ventilation and avoid breathing the fumes.

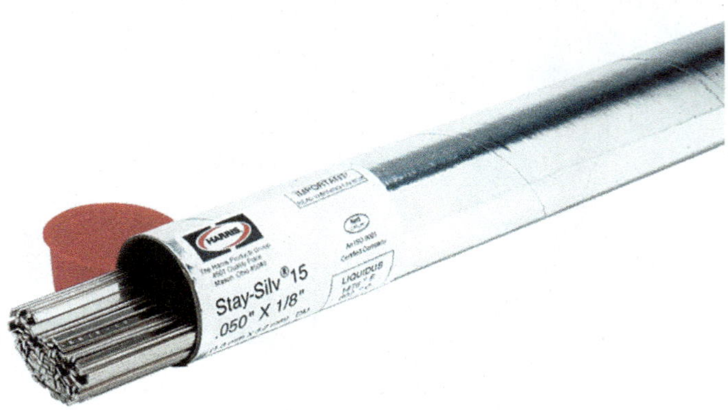

Figure 23 BCuP-5 brazing rods.
Source: Photo courtesy of The Lincoln Electric Company, Cleveland, OH, U.S.A.

Since brazing temperatures are significantly higher than soldering temperatures, oxides form even more rapidly. Without the proper use of flux when needed, the brazed joint will not reach an acceptable level of quality or strength. The properties of brazing fluxes are significantly different from soldering fluxes due to the temperatures involved. Brazing fluxes are more corrosive. Never mix the two types or substitute one for the other. For best results, use the flux recommended by the manufacturer of the brazing filler metals.

Brazing fluxes (*Figure 24*) are applied using the same methods and guidelines as soldering fluxes. High-temperature fluxes used in brazing are usually applied only to the end of the tube and not on the inside of the fitting.

Brazed joints often must be disassembled in the field for service purposes. Whenever a hermetic compressor is replaced, for example, a number of brazed joints usually must be taken apart. Flux should be applied to these joints as well before they are heated. Apply the flux on top of the visible filler metal and on the adjacent areas of the tube and fitting.

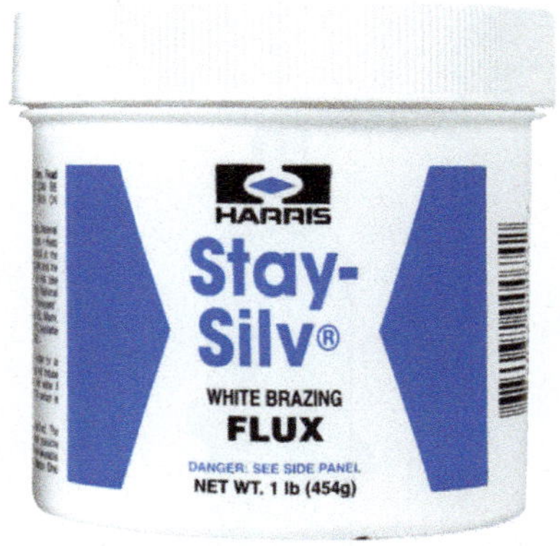

Figure 24 High performance brazing flux.

Source: Photo courtesy of The Lincoln Electric Company, Cleveland, OH, U.S.A.

2.2.0 Preparing for Brazing

The procedures for preparing tubing and fittings for brazing are similar to those for soldering. It is critical that proper cutting and cleaning techniques be used to produce a solid, leakproof joint. Use the following procedures to prepare tubing and fittings for brazing:

Step 1 Measure the distance between the faces of the two fittings.

Step 2 Determine the cup depth engagement of each of the fittings.

Step 3 Add the cup depth engagement of both fittings to the measurement found in *Step 1* to find the length of tubing needed.

Step 4 Cut the copper tubing to the correct length using a tubing cutter.

Step 5 Ream the inside and outside of both ends of the copper tubing using a reamer.

WARNING!

Do not touch or brush away filings from the tube or fitting with your bare fingers. Always use an appropriate tool. The ends of cut tubing, especially along the inside edge, are extremely sharp. In addition, natural oils in your fingers will contaminate the freshly cleaned metal.

Step 6 Clean the tubing and fitting using No. 00 steel wool, a piece of emery cloth, an abrasive pad, or a special copper-cleaning tool.

Step 7 Apply flux (if required) to the end of the copper tubing before inserting it into the fitting. Flux does not need to be applied to the fitting cup.

Step 8 Insert the tube into the fitting socket. Push and turn the tube into the socket until the tube touches the inside shoulder of the fitting.

Step 9 Wipe away any excess flux from the outside of the joint.

Step 10 Check the tube and fitting for proper alignment before brazing.

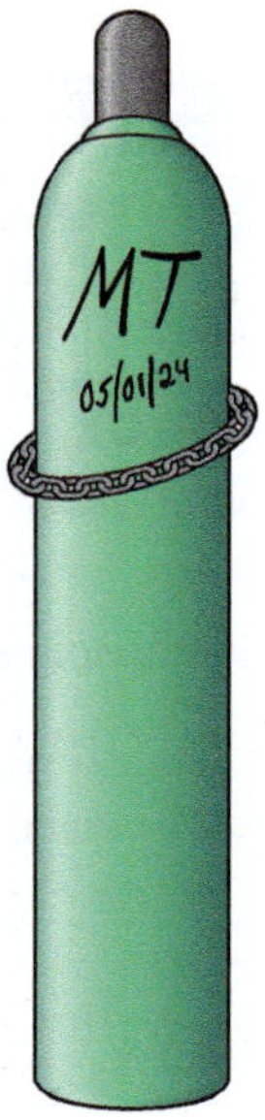

Figure 25 Typical empty cylinder marking.

2.2.1 Brazing Equipment Setup

The brazing procedure differs from soldering in that different equipment may be required to raise the temperature of the metals to be joined well above 800°F (427°C). Most brazed joints are made at temperatures between 1,200°F and 1,550°F (650°C and 840°C). This is the range of temperatures where the most popular brazing alloys are fluid enough to flow without becoming too fluid. The torch flame must be significantly higher than this range to achieve good control over the heating process. Because of the higher temperatures needed, oxyacetylene equipment is typically used for brazing, which will be further described in this section.

During use, transportation, and storage, oxygen and acetylene cylinders must be secured with a stout cable or chain to prevent them from falling, injuring people, or damaging equipment. At the jobsite, oxygen and acetylene cylinders must be stored separately with at least 20' (6 m) between them, or with a 5' (1.5 m) high, 30-minute minimum firewall separating them.

Cylinders should be stored without a regulator attached. The regulator should be removed any time a cylinder is moved. With a regulator attached, the chances of shearing off the cylinder valve stem significantly increase, since it protrudes well beyond the body of the cylinder.

Store empty cylinders away from full or partially full cylinders and make sure they are clearly marked to show that they are empty (*Figure 25*). Take care, however, to treat all oxygen and fuel gas cylinders as if they contain gas. Although they may be too empty to provide service, some gas will likely remain.

Alternate High-Pressure Cylinder Valve Cap

High-pressure cylinders can be equipped with a clamshell cap that can be closed to protect the cylinder valve with or without a regulator installed on the valve. This enables safe movement of the cylinder after the cylinder valve is closed. This type of cap is usually secured to the cylinder body cap threads when it is installed so that it cannot be removed. When the clamshell is closed, it can also be padlocked to prevent unauthorized operation of the cylinder valve.

(A) Clamshell Open to Allow Cylinder Valve Operation

(B) Clamshell Closed for Movement or Padlocked to Prevent Operation of Cylinder Valve

(C) Clamshell Closed for Transport

Alternate Acetylene Cylinder Safety Cap

Acetylene cylinders can be equipped with a ring guard cap that protects the cylinder valve with or without a regulator installed on the valve. This enables safe movement of the cylinder after the cylinder valve is closed. This type of cap is usually secured to the cylinder body cap threads when it is installed so that it cannot be removed.

(A)

(B)

Follow this basic procedure to prepare oxyacetylene brazing equipment for use:

WARNING!

Do not handle acetylene and oxygen cylinders with oily hands or greasy gloves. Keep grease away from the cylinders and do not use oil or grease on cylinder attachments or valves. The mixture of oil and oxygen could cause an explosion. Make sure that the protective caps are in place on the cylinders before transporting or storing them.

Step 1 Install and securely fasten the oxygen and acetylene cylinders in a bottle cart or in an upright position against a wall or other substantial structure.

WARNING!

Do not allow anyone to stand in front of the oxygen cylinder valve when opening (cracking) it. The oxygen is under high pressure (about 2,000 psig or 140,000 kPa) and could blow debris into eyes or face. In addition, the bottle reaction could cause severe injury if the valve is opened unintentionally. Ensure that the bottle is properly secured before cracking the valve open.

Step 2 Install the oxygen regulator on the oxygen cylinder as follows:

- Remove the cylinder protective cap.
- Open (crack) the oxygen cylinder valve just long enough to allow a small amount of oxygen to pass through and blow out any debris, then close it.
- Carefully inspect the regulator attaching threads, cylinder threads, and mating surfaces for damage (*Figure 26*).
- Turn the adjusting screw on the oxygen regulator counterclockwise (out) until it is loose. This will shut off the regulator output and prevent accidental overpressurizing of the hose and torch during hookup.
- Using a torch wrench, install the oxygen regulator on the cylinder (*Figure 27*). Oxygen cylinders and regulators have right-handed threads. Tighten the nut snugly. Be careful not to overtighten the nut because this may strip the threads.

WARNING!

Acetylene gas is flammable. Always inspect the surrounding area to ensure there are no open flames or sparks present.

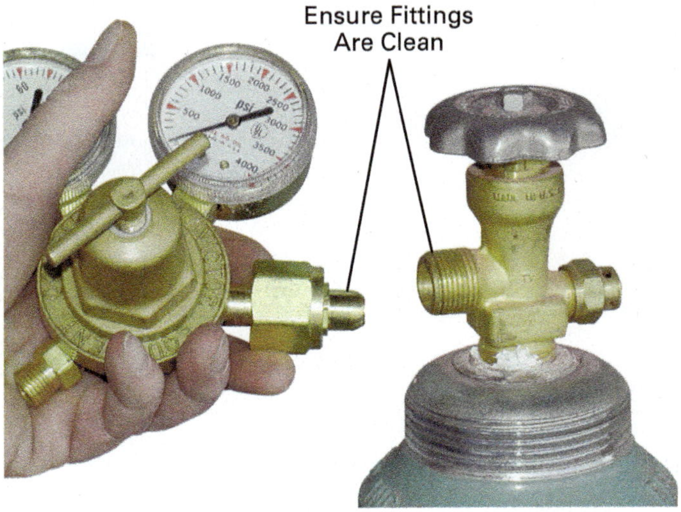

Figure 26 Inspecting the fittings.

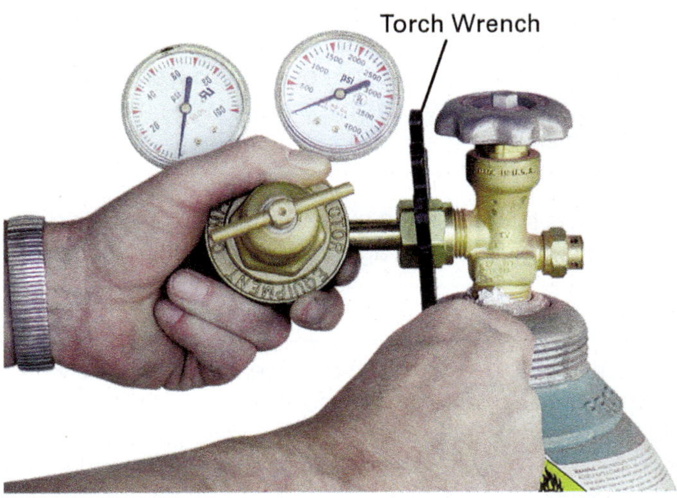

Figure 27 Installing the regulator.

Step 3 Install the acetylene regulator on the acetylene cylinder as follows:

- First ensure that the acetylene usage rate for the task does not exceed the recommended safe usage rate for the cylinder. For example, if the cylinder holds 40 ft^3 (1.13 m^3) of acetylene, the usage rate for intermittent use should not exceed 10% of the volume, or 4 ft^3 (0.11 m^3), per hour.
- Remove the cylinder protective cap.
- Using the valve wrench, open (crack) the acetylene cylinder valve just long enough to allow a small amount of acetylene to pass through the valve. Then close it.

- Turn the adjusting screw on the acetylene regulator counterclockwise until it is loose. This will shut off the regulator output and prevent accidental overpressurizing of the hose and torch during hookup.
- Using a torch wrench, install the acetylene regulator on the cylinder. Acetylene cylinders and regulators have left-handed threads. Tighten the nut snugly. Be careful not to overtighten the nut because this may strip the threads.

Step 4 Connect the hoses and brazing torch as follows:

- If they are not already in place or a part of the torch handle, install flashback arrestors on the oxygen and acetylene hoses at the torch handle inlets.
- Connect the green hose to the oxygen regulator and the red hose to the acetylene regulator. Tighten the hoses snugly. Be careful not to overtighten the fittings because this may strip the threads.

WARNING!

Do not allow anyone to stand to the front or rear of the oxygen regulator when opening the cylinder valve in case the regulator fails suddenly. Open the oxygen cylinder valve slowly.

Step 5 Purge (clean) the oxygen hose as follows:

- Crack the oxygen cylinder valve open slowly until the pressure stops rising on the cylinder gauge. This now indicates the pressure in the cylinder. Then slowly open the valve completely.
- Turn the oxygen regulator adjusting screw clockwise (in) until a small amount of pressure (3 psig to 5 psig, or 20 kPa to 35 kPa) shows on the oxygen working-pressure gauge. Allow a small amount of pressure to build up and purge the oxygen hose, clearing it of any loose debris.
- Turn the oxygen regulator adjusting screw counterclockwise (out) until it is again loose. This will shut off the regulator output.

Step 6 Purge (clean) the acetylene hose as follows:

- Crack the acetylene cylinder valve open slowly until the cylinder pressure registers on the acetylene cylinder gauge. Then open it about $\frac{1}{2}$ to $\frac{3}{4}$ turn.
- Turn the acetylene regulator adjusting screw clockwise (in) until a small amount of pressure shows on the acetylene working-pressure gauge. Allow a small amount of pressure to build up and purge the acetylene hose, clearing it of any loose debris.
- Turn the acetylene regulator adjusting screw counterclockwise (out) until it is again loose. This will shut off the regulator output.

Step 7 Install the brazing torch on the ends of the hoses and close the valves on the torch. Remember that the oxygen hose has normal right-handed threads while the acetylene hose has left-handed threads.

WARNING!

Never adjust the acetylene regulator to a pressure higher than 15 psig (103 kPa). Above this pressure, acetylene decomposes, becoming unstable and volatile.

Step 8 Check the oxyacetylene equipment for leaks as follows:

- Adjust the acetylene regulator adjusting screw for the appropriate pressure on the working-pressure gauge, as recommended by the torch tip manufacturer for the application.
- Adjust the oxygen regulator adjusting screw for the appropriate pressure on the working-pressure gauge, as recommended by the torch tip manufacturer for the application.
- Close the oxygen and acetylene cylinder valves and check for leaks while the pressure remains in the hoses and torch handle. If the working-pressure gauges remain at the set pressures for several minutes, there should be no significant leaks in the hose system. If the readings drop, there is a leak. Use a soap solution or commercial leak detection fluid to check the oxygen or acetylene connections for leaks. Note that the leak could also be on the regulator assembly. If the regulator assembly is found to be leaking, stop using it immediately. Do not attempt to repair it in the field.

WARNING!

Do not use an oil-based soap for leak testing because the mixture of oil and oxygen may cause an explosion.

- Open both valves on the torch handle to release the pressure in the hoses. Watch the working-pressure gauges until they register zero, then close the valves on the torch handle.
- Turn the oxygen and acetylene regulator valves counterclockwise until they are loose. This will release the spring pressure on the regulator diaphragms and completely close the regulators.

Step 9 Coil the hoses and hang them on the hose holder.

2.2.2 Lighting the Oxyacetylene Torch

After the oxyacetylene brazing equipment has been properly prepared, the torch can be lit and the flame adjusted for brazing. Before beginning, it is important to know the correct oxygen and acetylene operating pressures for the tip in use. *Table 6* provides one example of a manufacturer's table with this data for a particular series of tips. Note that this table is designed for the manufacturer's products in North America. Therefore, this example does not include metric values. This table should not be considered a general guide for all tips. Consult the information for the actual tip in use.

TABLE 6 Brazing Tip Operational Data

Part #	UPC #	Acetylene Consumption (SCFH)	Fuel Pressure (psig)	Oxygen Pressure (psig)	Welding Metal Thickness	Brazing Copper Tubing
Type 17-000	30250	1–2	5	5	1/32"	1/8"
Type 17-00	30251	2–3	5	5	3/64"	1/4"
Type 17-0	30252	2–4	5	5	5/64"	1/2"
Type 17-1	30253	3–6	5	5	3/32"	3/4"
Type 17-2	30254	5–10	5	5	1/8"	1"
Type 17-3	30255	8–18	6	7	3/16"	1 1/2"
Type 17-4	30256	10–25	7	10	1/4"	2"
Type 17-5	30257	15–35	8	12	1/4" to 1/2"	3"
Type 17-6	30258	25–45	9	14	1/2" to 3/4"	4"
Type 17-7	30259	30–60	10	16	3/4" to 1 1/4"	6"

For hose lengths greater than 25' (8 m), the pressure at the regulator is generally increased 2 psig to 3 psig (14 kPa to 20 kPa) to compensate for pressure drop. Remember that the acetylene pressure should never exceed 15 psig (103 kPa) at any point beyond the cylinder.

There are three types of flames: carburizing (reducing), neutral, and oxidizing (*Figure 28*).

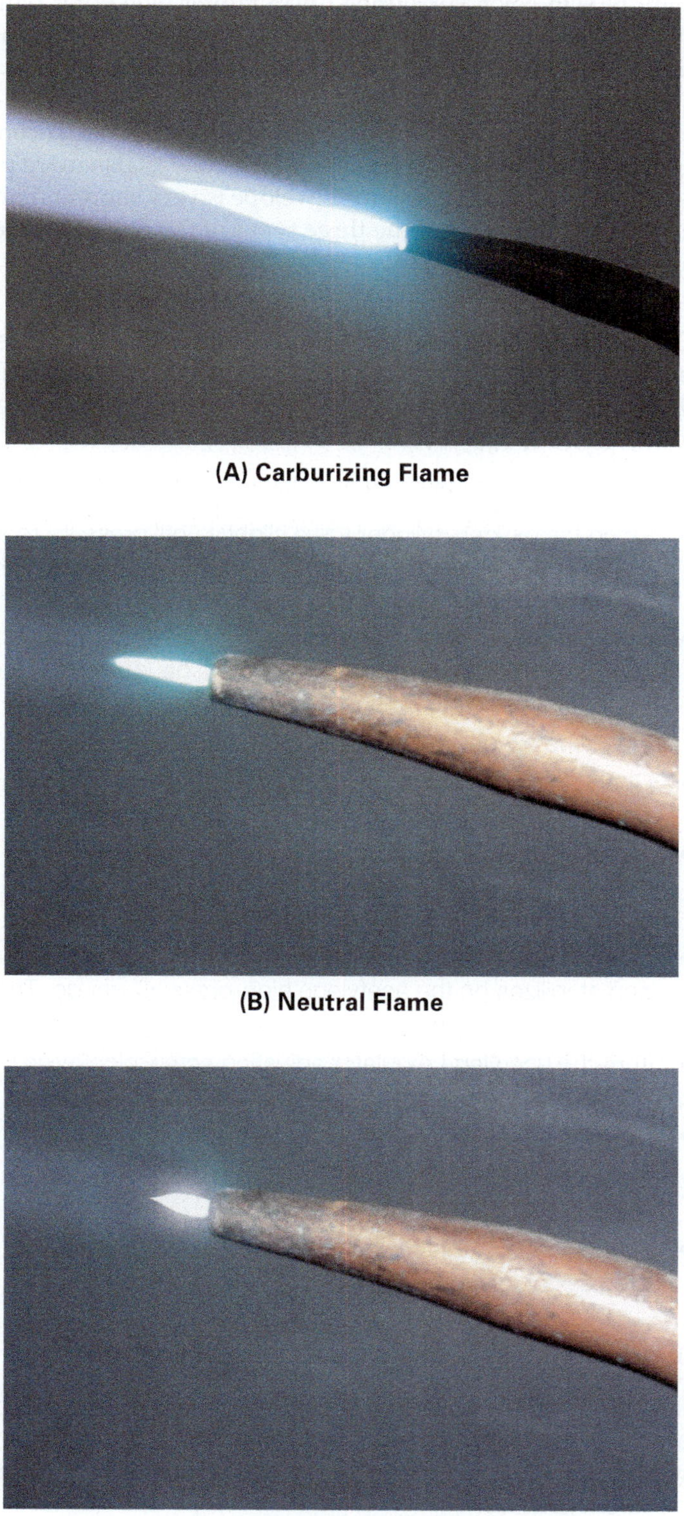

(A) Carburizing Flame

(B) Neutral Flame

(C) Oxidizing Flame

Figure 28 Types of flames.

A carburizing (reducing) flame has a white feather created by excess fuel. The length of the feather depends on the amount of excess fuel in the flame. The outer flame envelope is brighter than that of a neutral flame and much lighter in color. The excess fuel in the carburizing flame produces large amounts of carbon. The carburizing flame is cooler than the neutral flame, and it is used for light brazing to prevent melting of the base metal.

The neutral flame burns equal amounts of oxygen and acetylene. The inner cone is bright blue in color, surrounded by a fainter blue outer flame envelope that results when the oxygen in the air combines with the superheated gases from the inner cone. A neutral flame is used for almost all fusion welding or heavy brazing applications.

An oxidizing flame has an excess amount of oxygen. Its inner cone is shorter, with a bright blue edge and a lighter center. The cone is also more pointed than that of a neutral flame. The outer flame envelope is very short and often fans out at the ends. This is the hottest of the three types of flames. It is sometimes used for brazing cast iron or other metals.

Use the following procedure to light an oxyacetylene torch:

Step 1 Set up the oxyacetylene torch, following all safety procedures discussed earlier. Install the appropriate tip before lighting the torch. Refer to *Table 4* for recommended tip sizes. Be aware of the gas pressure settings required for the tip in use.

Step 2 Adjust the torch oxygen flow as follows:

- Open the oxygen cylinder valve slightly until pressure registers on the oxygen cylinder high-pressure gauge and then stops rising. Then open the valve fully.
- Turn the oxygen regulator adjusting screw clockwise until pressure shows on the oxygen working-pressure gauge. Adjust the regulator for the manufacturer's recommended pressure setting.
- Open the oxygen valve on the torch handle.
- Check the regulator and ensure that the pressure remains at the desired setting. Adjust as needed.
- Close the oxygen valve on the torch handle.

NOTE

Always fine-tune the pressure setting with the torch valve open. The pressure may register slightly higher when it is closed. Regulators are incapable of making pressure adjustments without flow.

Step 3 Adjust the torch acetylene flow as follows:

- Open the acetylene cylinder valve slightly until the pressure registers and stabilizes on the acetylene high-pressure gauge. Then open the valve about $\frac{1}{2}$ turn.
- Turn the acetylene regulator adjusting screw clockwise until the desired pressure shows on the acetylene working-pressure gauge.
- Open the acetylene valve on the torch handle.
- Check the regulator and ensure that the pressure remains at the desired setting. Adjust as needed.
- First close, and then open the torch acetylene valve about $\frac{1}{2}$ turn to prepare for ignition.

CAUTION

Be sure to leave the valve wrench on the acetylene cylinder valve while the cylinder is in use so that the valve can be closed quickly in case of an emergency.

WARNING!

When lighting the torch, be sure to wear flame-resistant gloves and goggles. Use an appropriate device, such as a cup-type striker, to light the torch, and point the torch away from you. Always light the fuel gas first, and then open the oxygen valve on the torch handle.

Step 4 Light the torch as follows:

- Hold the striker in one hand and the torch in your other hand. Strike a spark in front of the escaping acetylene gas.
- Open the acetylene valve on the torch until the flame jumps away from the tip about $\frac{1}{16}$" (1.6 mm). Then slowly close the valve until the flame just returns to the tip (*Figure 29*). This sets the proper fuel gas flow rate for the size tip being used.
- Open the oxygen valve on the torch slowly to add oxygen to the burning acetylene. Refer again to *Figure 28* and *Figure 29*. Observe the luminous cone at the tip of the nozzle and the long, greenish envelope around the flame, which is excess acetylene representing a carburizing flame. As you continue to add oxygen, the envelope of acetylene should disappear. The inner cone will appear soft and luminous, and the torch will make a soft, even, blowing sound. This indicates a neutral flame, which is the ideal flame for brazing. If too much oxygen is added, the flame will become smaller, more pointed, and white in color, and the torch will make a sharp snapping or whistling sound. For brazing thin materials, a cooler, carburizing flame can be used to help prevent melting the base metal accidentally.

NOTE

It is not unusual for visible soot to form in strings and float away from the acetylene-only flame. They can fall on an otherwise clean surface and be somewhat messy. Be aware that soot is likely to form from the initial flame and be aware of your surroundings as you light the torch.

Step 5 Shut off the torch when finished brazing as follows:

- Shut off the oxygen valve on the torch handle first.
- Shut off the acetylene valve on the torch handle. Close it quickly to avoid excess carbon buildup.
- Shut off both the oxygen and acetylene cylinder valves completely.
- Open the acetylene valve on the torch handle to release the pressure in the hose while watching the gauge. With the cylinder valve closed, the pressure should fall on the regulator gauges until they indicate zero pressure. Then close the valve on the torch handle. Repeat the procedure for the oxygen delivery system.
- Turn the oxygen and acetylene regulator adjusting knobs counterclockwise (out) until loose. This will release the spring pressure on the diaphragms in the regulators.

Step 6 Properly stow the hoses.

(A) Initial Flame

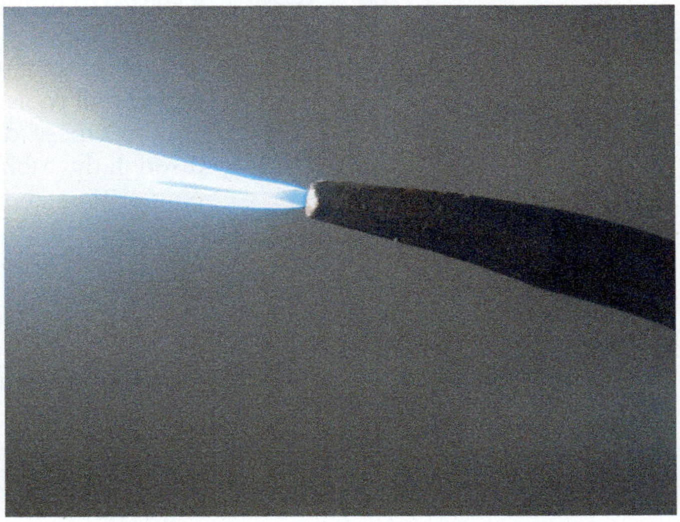

(B) Final Flame Adjustment

Figure 29 Reduce the acetylene flow until the flame returns to the tip.

Cylinder Identification Transducers and Safety Plugs

Many gas suppliers mount transducers on their cylinders so that they can readily identify them. The transducer is electronically scanned, and a coded number is matched against the supplier's records for identification. This aids in quickly determining the purchaser or user of the cylinder as well as the required retesting date. Acetylene cylinders are also equipped with safety plugs that will release if the temperature exceeds 220°F (104°C). The safety plugs release the gas in the event of a fire, thereby preventing the cylinder from exploding.

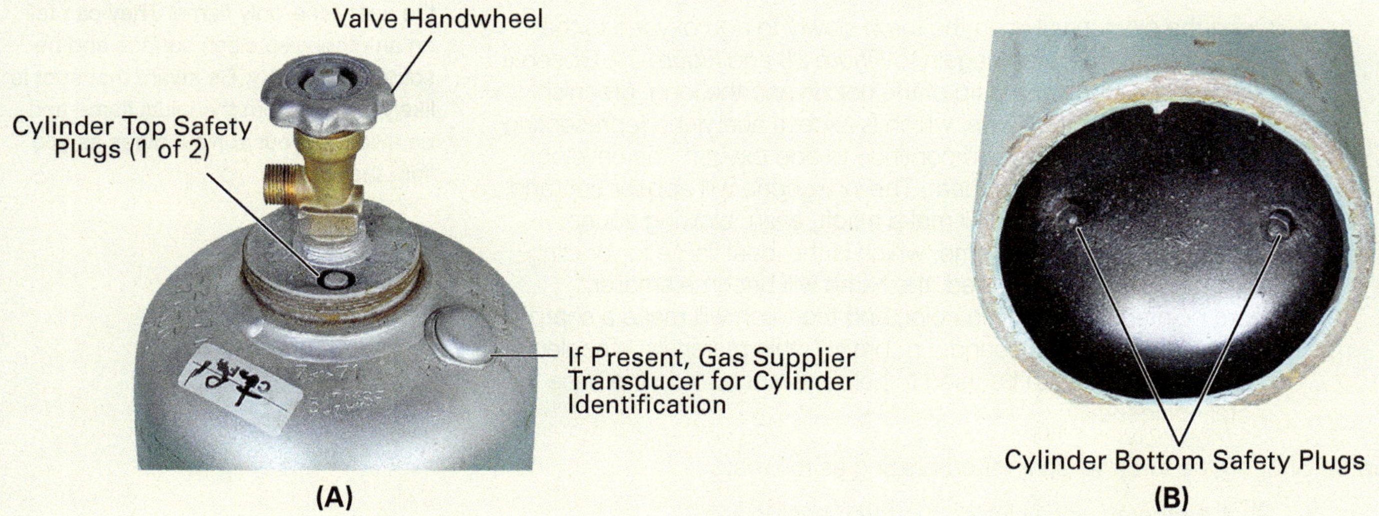

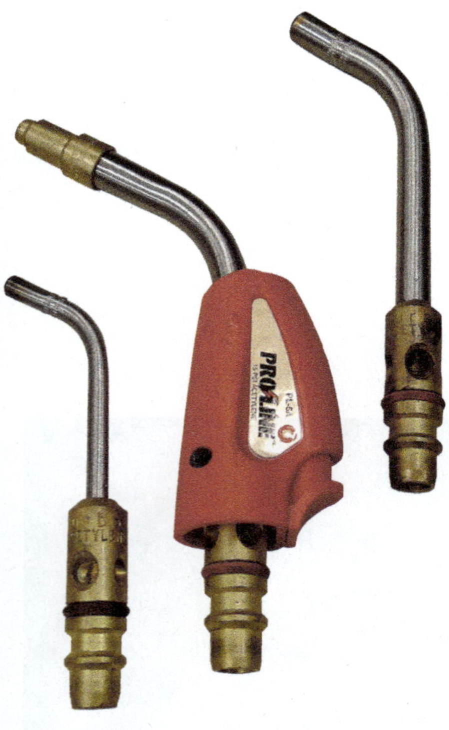

Figure 30 Air-acetylene torch tips.

2.2.3 Air-Acetylene Equipment Setup

The heat provided by an air-acetylene torch alone is often suitable for brazing smaller lines. The procedure for setting up and using the air-acetylene torch is the same as that used in the soldering process.

Air-acetylene torch tips are available in different sizes (*Figure 30*). Note that a larger tip does not increase the temperature of the flame, but it provides more heat through increased coverage of the material. Follow the manufacturer's guidelines for the selection of torch tips.

2.2.4 Purging Lines

Oil inside tubing or a part being brazed can vaporize when the heat of the brazing torch is applied. Oil vapor mixed with air can catch fire or explode if a source of ignition is added. When copper is heated during brazing, it reacts rapidly with the oxygen in the air to form copper oxide. The oxidation occurs in such abundance that it forms thin, flaky deposits that separate from the tubing as it cools. If oxygen from the atmosphere is present in the tubing during brazing, these deposits form inside the tubing as well. Liquid or gas flow through the tubing will later dislodge the copper oxide particles, which tend to plug orifices and generally pollute the system. Therefore, all the air should be removed from the tubing being brazed. This can be best done by purging the tubing with nitrogen.

WARNING!

Never use oxygen to purge tubing. An explosion could result when oil and oxygen are mixed.

The pressure in a full nitrogen cylinder is 2,000 psig (140,000 kPa) or more. An accurate nitrogen pressure regulator and an adjustable pressure relief valve must always be used when purging (*Figure 31*). The relief valve should be adjusted to open at a relatively low pressure, such as 7 psig to 10 psig (50 kPa to 70 kPa).

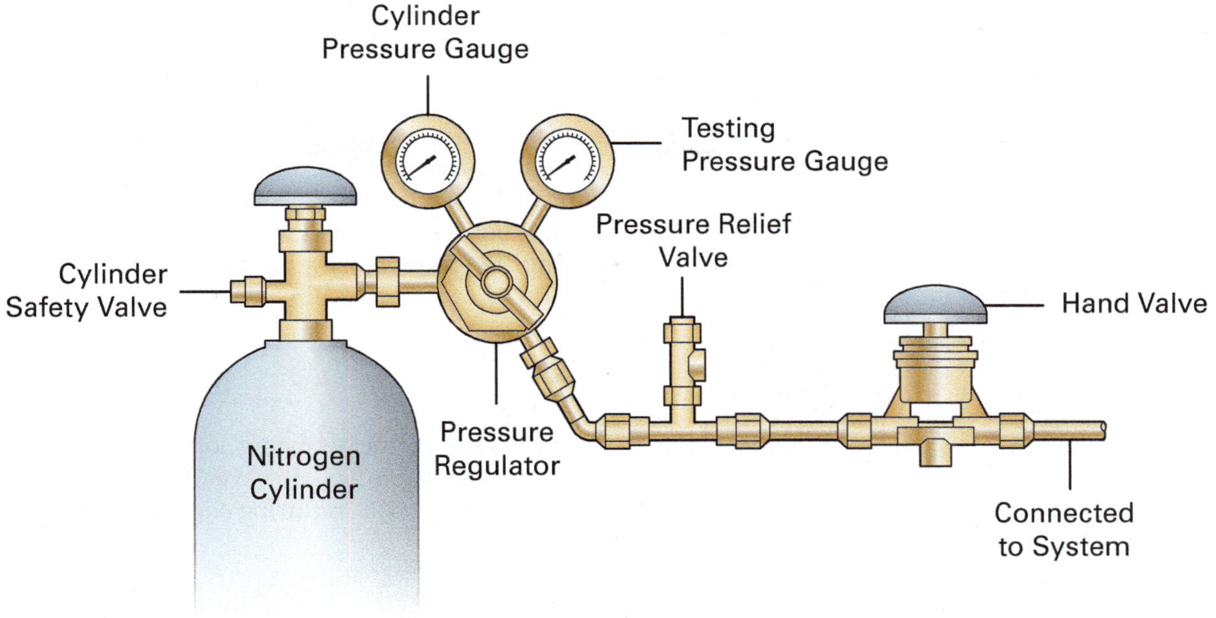

Figure 31 Nitrogen purging setup.

The goal of purging a system for brazing is to simply push the air (and the oxygen it contains) out of the tubing and keep it out while the brazing is being done. However, pressure cannot be allowed to build in the tubing while brazing proceeds. Even a small positive pressure will allow leaks to develop as nitrogen escapes through a joint while the filler metal is in its liquid state. For this reason, the tubing circuit must remain freely open to the atmosphere at one or more points, allowing the minimal flow of nitrogen to escape without a buildup of pressure. Purging does not require the use of any significant pressure. Once the air is initially pushed out, the goal is to simply create and maintain a nitrogen atmosphere inside the tubing. The nitrogen pressure inside the tubing, therefore, needs only to be slightly higher than the atmospheric pressure.

The best approach is to connect the nitrogen output from the regulator to a gauge manifold. A gauge manifold is simply a gauge combined with a needle valve or similar hand-operated valve to fine-tune the nitrogen flow. A needle valve provides much finer control over the nitrogen flow than is possible with the nitrogen regulator alone. The manifold can also be equipped with multiple hoses that can be connected to various points of the system. A refrigeration gauge manifold works very well for this purpose.

When brazing independent tubing assemblies or circuits that have no gauge ports, ports can be installed or cone valve plugs (*Figure 32*) can be used. The plugs are pushed into the end of the tubing or fitting to provide a point of connection.

Once the connections are made and the regulator is installed, slowly open the nitrogen cylinder valve. With the gauge manifold valves closed, set the regulator for roughly 3 psig to 5 psig (20 kPa to 35 kPa). Then open the gauge manifold valve(s) and allow the nitrogen to begin circulating through the system. Once the air has been pushed out, use the gauge manifold valve to reduce the flow to a very low volume—just enough to keep the air from flowing back in. The flow of nitrogen is generally sufficient when the flow can be felt gently in the palm of your hand. Allow the flow to continue at this level throughout the brazing process and as the joint cools.

If a tubing circuit is being closed and the final joints are being brazed, it is essential that a port remain open near the work to relieve any pressure. As the last few joints are completed, it is often best to turn off the nitrogen flow altogether. Heat from the process will typically maintain enough of a draft leaving the circuit to prevent a significant amount of air from entering.

Figure 32 Cone valve plug.
Source: Uniweld Products, Inc. "Quality Tools That Go To Work With You!"

When tubing brazing takes place in a confined space, make sure to attach a hose or pipe to the tubing being brazed so that the purge gas leaving the tubing is vented to the atmosphere outside. Otherwise, a hazardous atmosphere can develop within the space as the oxygen is displaced, resulting in an oxygen deficiency. OSHA regulations require that the internal atmosphere of a permit-required confined space be tested for hazardous atmosphere with a calibrated direct-reading instrument before anyone is allowed to enter the space. The atmosphere must be continuously monitored while work is in progress to detect any conditions indicating the presence of a hazardous atmosphere, including oxygen deficiency.

2.3.0 Brazing Joints

When brazing joints, the types of metals being joined must be considered. Copper-to-copper joints are very common. However, there will be times when dissimilar metals may need to be joined, requiring slightly different techniques.

2.3.1 Brazing Copper to Copper

When brazing copper to copper, first conduct an initial assessment of the work area to determine what is required to complete the brazing tasks. Obtain all the materials and equipment needed to properly set up and begin the brazing process. Once the setup is complete and the flame is properly adjusted, the torch is ready to braze joints.

Use the following procedure to set up and braze a copper-to-copper joint:

Step 1 Prepare the work area and protect any components in the piping system that could be damaged by the heat from the brazing process. Remove any loose, flammable materials in the area and ensure there is adequate room for physical movement around the joint. Place an appropriate fire extinguisher within easy reach.

Step 2 Set up the nitrogen gas to purge the tubing, following the guidelines and precautions described previously.

Step 3 Put on safety glasses or welding goggles along with a face shield, one of which should be tinted with a #3 or #4 shade.

Step 4 Set up and light the oxyacetylene brazing equipment as described previously:

- Place the chosen filler metal within easy reach and be certain that the material is freely accessible while the brazing is in progress.
- Light and adjust the torch to produce a neutral flame.

Step 5 Apply the heat to the tubing first in the area adjacent to the fitting. Watch the flux, if used. It will bubble, turn white, and then melt into a clear liquid, eventually becoming calm. At this time, shift the flame to the fitting and hold it there until any flux on the fitting turns clear.

Step 6 Continue to move the heat back and forth over the tubing and the fitting. Allow the fitting to receive more heat than the tubing by briefly pausing at the fitting while continuing to move the flame back and forth. Pause at the back of the fitting cup. For $1\frac{1}{4}$" (3.175 cm) and larger tubing, it may be difficult to bring the whole joint up to temperature at one time with a single-orifice tip. It is often better to use a multiflame (rosebud) heating tip, such as the one shown in *Figure 33*, to maintain a more uniform temperature over large areas. A mild preheating of the entire fitting and adjacent tubing is recommended for larger sizes. The use of a second torch to retain a uniform preheating of the entire fitting assembly may be necessary with very large diameters.

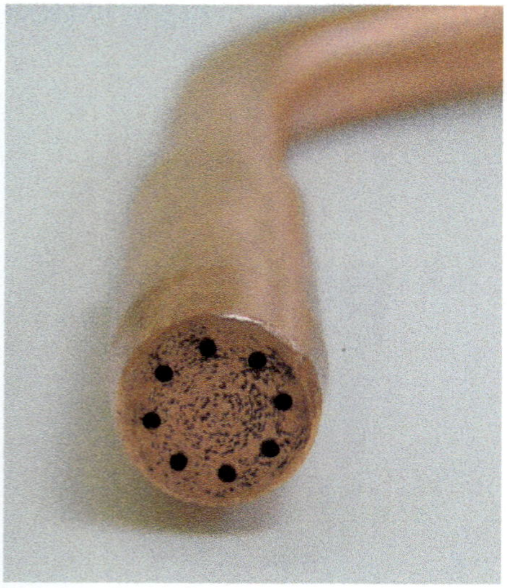

Figure 33 Multiflame heating tip.

Step 7 Touch the filler metal rod to the joint. If the filler metal does not melt on contact, continue to heat and test the joint until the filler metal melts. Be careful to avoid melting the base metal.

Step 8 Hold the filler metal rod to the joint and allow the filler metal to enter into the joint. At the same time, hold the torch slightly ahead of the filler metal and direct most of the heat to the shoulder of the fitting. When the filler metal is applied to the top of the fitting, it should run down to the bottom and fill in the joint through capillary action. Applying heat at the bottom of the joint should help to make sure the filler metal penetrates. Make sure the filler metal is visible all around the shoulder of the joint. If it is not, apply additional filler metal. However, building an excessive fillet does not improve joint quality and wastes material.

Step 9 For larger joints, small sections of the joint can be heated and brazed. Be sure to overlap the previously brazed section as you continue around the fitting, as shown in *Figure 34*.

Step 10 After the filler metal has hardened and while it is still hot, wipe the joint with a wet rag to clean off any excess brazing flux. This also helps the copper oxides to crack and break away in flakes.

Step 11 Allow the joint to finish cooling naturally. Maintain the nitrogen flow until the joint is cool.

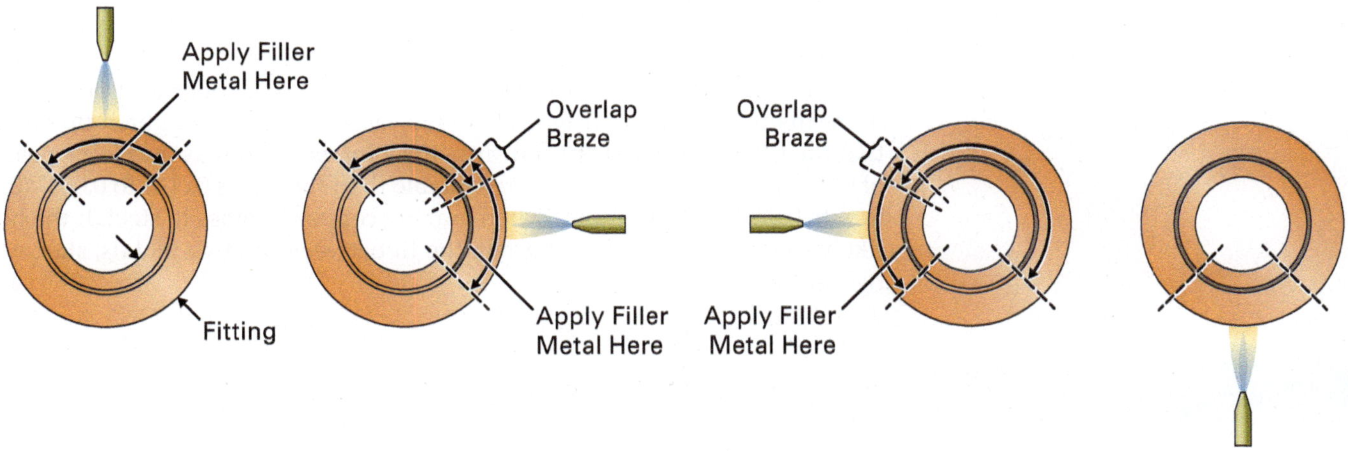

Figure 34 Working in overlapping sectors.

The technique for brazing will vary somewhat, depending on the position in which the brazing is being done. The method just described is good for horizontal runs of pipe, but some adjustments are needed for vertical-up and vertical-down brazing:

- *Vertical-up joints* — Heat the tube first, and then transfer the heat of the torch to the fitting. Move the heat back and forth between the fitting and the tube but be sure not to overheat the tube below the fitting. Doing so could cause the filler metal to flow out of the joint. When the brazing temperature is reached, apply the filler metal to the joint while applying heat to the wall of the fitting. The heat and capillary action should cause the filler metal to run up into the fitting.
- *Vertical-down joints* — Heat the tube before heating the fitting. When the brazing temperature is reached, apply more heat to the fitting while applying the filler metal to the joint. The filler metal should run down into the fitting.

Figure 35 A 45% silver brazing alloy in roll form (BAg-5).
Source: Courtesy of Harris Products Group

2.3.2 Brazing Dissimilar Metals

Some assemblies and subassemblies used in various types of system equipment or processes have fittings that are made of steel, copper-clad steel, or brass. Refrigeration system accumulators and receivers, for example, often have steel fittings, while some refrigeration filter-driers have copper-clad steel fittings. Other components may be made of brass. Special attention must be given to these types of assemblies.

A 45% silver brazing alloy meeting BAg-5 specifications (*Figure 35*), along with an appropriate flux, is usually a good choice for brazing dissimilar metals such as copper to brass or copper to steel. A non-phosphorous alloy is generally required because phosphorous could cause the joint to become brittle and fail. BAg-5 products do not contain phosphorus. Remember that, for copper-to-copper joints, a phosphorous-bearing filler metal provides the flux, so no additional flux is required. When a non-phosphorous filler metal is used, a flux must be used, even on copper. As always, the pieces being joined must be very clean and free from any oil or grease.

Aside from choosing the correct filler metal and flux, the other important aspect of brazing dissimilar metal is the application of heat. The fittings on accessories that use steel or copper-clad steel fittings reach their melting point faster than copper, so there is a potential risk of overheating and distorting the fittings during brazing. To avoid overheating, it is necessary to move the heat rapidly and uniformly around the joint and not let it linger too long in one place. Thin steel can turn cherry red very quickly; because it transfers heat away much more slowly than copper, its temperature rises rapidly. When joining copper to steel, more heat needs to be applied to the copper, since is the better conductor and will dissipate heat more quickly than the steel.

Remember that metals expand at different rates. Copper and brass expand more than steel when heated. This may cause the joint to tighten up so much that the filler metal will not flow into the joint. For that reason, a joint clearance of 0.010" (0.254 mm) may be more suitable than the usual 0.001" to 0.003" (0.0254 mm to 0.0762 mm) clearance when brazing copper or brass to steel. If the heat is likely to loosen rather than tighten the fit between the two metals, start with a tighter fit.

2.0.0 Section Review

1. An acetylene cylinder that has been laid down should be placed in the upright position for at *least* _____.
   a. 30 minutes before use
   b. 1 hour before use
   c. 2 hours before use
   d. 4 hours before use

2. Oxygen cylinders are typically pressurized to a pressure of about _____.
   a. 15 psig (103 kPa)
   b. 250 psig (1,725 kPa)
   c. 500 psig (3,450 kPa)
   d. 2,000 psig (140,000 kPa)

3. Which of the following filler metals is *best* used for brazing dissimilar metals?
   a. A BCuP-3 filler metal
   b. A BAg-5 filler metal
   c. A BCuP-5 filler metal
   d. Any phosphorous-bearing alloy

Module 29405 Review Questions

1. The filler metals commonly used for soldering usually melt in a temperature range of _____.
   a. 200°F to 375°F (93°C to 191°C)
   b. 375°F to 500°F (191°C to 260°C)
   c. 750°F to 850°F (400°C to 454°C)
   d. 1,000°F to 1,200°F (538°C to 650°C)

2. The volume capacity of an acetylene B cylinder is _____.
   a. 10 ft^3 (0.28 m^3)
   b. 20 ft^3 (0.57 m^3)
   c. 30 ft^3 (0.85 m^3)
   d. 40 ft^3 (1.13 m^3)

3. Which of the following is an *incorrect* statement regarding the filler metal used for soldering?
   a. It is a nonferrous metal.
   b. It is a metal alloy.
   c. It has a melting point below 800°F (427°C).
   d. It always contains some percentage of lead.

4. The measured length of a pipe should include the distance between the fitting faces plus _____.
   a. the cup depth of the two fittings
   b. ½ the total length of the two fittings
   c. the distance to the centers of the two fittings
   d. approximately ½" (12.7 mm) on either end of the pipe

5. Acetylene becomes volatile when the pressure setting is above _____.
   a. 5 psig (35 kPa)
   b. 7 psig (50 kPa)
   c. 10 psig (70 kPa)
   d. 15 psig (103 kPa)

6. The desired joint clearance for a brazed copper-to-copper joint is _____.
   a. 0.001" to 0.003" (0.025 mm to 0.08 mm)
   b. 0.003" to 0.005" (0.08 mm to 0.13 mm)
   c. approximately 0.015" (0.38 mm)
   d. 5% of the pipe diameter

7. To produce a hotter flame, oxyacetylene equipment mixes acetylene, as a fuel gas, with pure _____.
   a. nitrogen
   b. oxygen
   c. argon
   d. helium

8. The type of brazing filler metal that contains cadmium is _____.
   a. BCuP-1
   b. BCuP-2
   c. BAg-2
   d. BAg-5

9. The filler metals commonly used for brazing melt at a temperature range of _____.
   a. 375°F to 500°F (191°C to 260°C)
   b. 750°F to 850°F (400°C to 454°C)
   c. 850°F to 1,200°F (454°C to 650°C)
   d. 1,200°F to 1,550°F (650°C to 840°C)

10. The type of flame used for almost all heavy brazing applications is a(n) _____.
    a. oxidizing flame
    b. carburizing flame
    c. neutral flame
    d. feathering flame

Answers to odd-numbered questions are found in the Review Question Answer Keys at the back of this book.

Answers to Section Review Questions

Answer	Section Reference	Objective
Section 1.0.0		
1. b	1.1.1	1a
2. c	1.2.0	1b
3. d	1.3.0	1c
Section 2.0.0		
1. a	2.1.1	2a
2. d	2.2.1	2b
3. b	2.3.2	2c

User Update

SCAN ME

Did you find an error? Submit a correction by visiting **https://www.nccer.org/olf** or by scanning the QR code using your mobile device.

APPENDIX 29401A

Performance Accreditation Tasks

Performance Accreditation Tasks (PATs) provide specific acceptable criteria for performance and help to ensure a true competency-based welding program for students.

PATs 4, 5, 7, 8, and 9 have no AWS correlation. PATs 1, 2, 3, and 6 correspond to and support learning objectives in *AWS EG3.0, Guide for the Training of Welding Personnel: SENSE Level II—Advanced Welders*. PAT 1 also corresponds to *AWS QC10, Specification for Qualification and Certification of SENSE Level I—Entry Welders*.

Each of the tasks are designed to evaluate your ability to run stringer beads, weave beads, and overlapping beads. In addition, fillet and V-groove welds must be completed with GMAW equipment in four standard test positions using aluminum filler wire of the appropriate diameter and shielding gas. Perform each task when you are instructed to do so by your instructor. As you complete each task, show it to your instructor for evaluation. Do not proceed to the next task until told to do so by your instructor.

Performance Accreditation Tasks

Module 29401

Weld A Pad On Aluminum Plate in the Flat (1G) Position Using GMAW Stringer Beads

As directed by the instructor, use GMAW process with the appropriate aluminum filler wire to make the following welds on aluminum plate: stringer beads, weave beads, weld restarts, weld terminations, and overlapping beads.

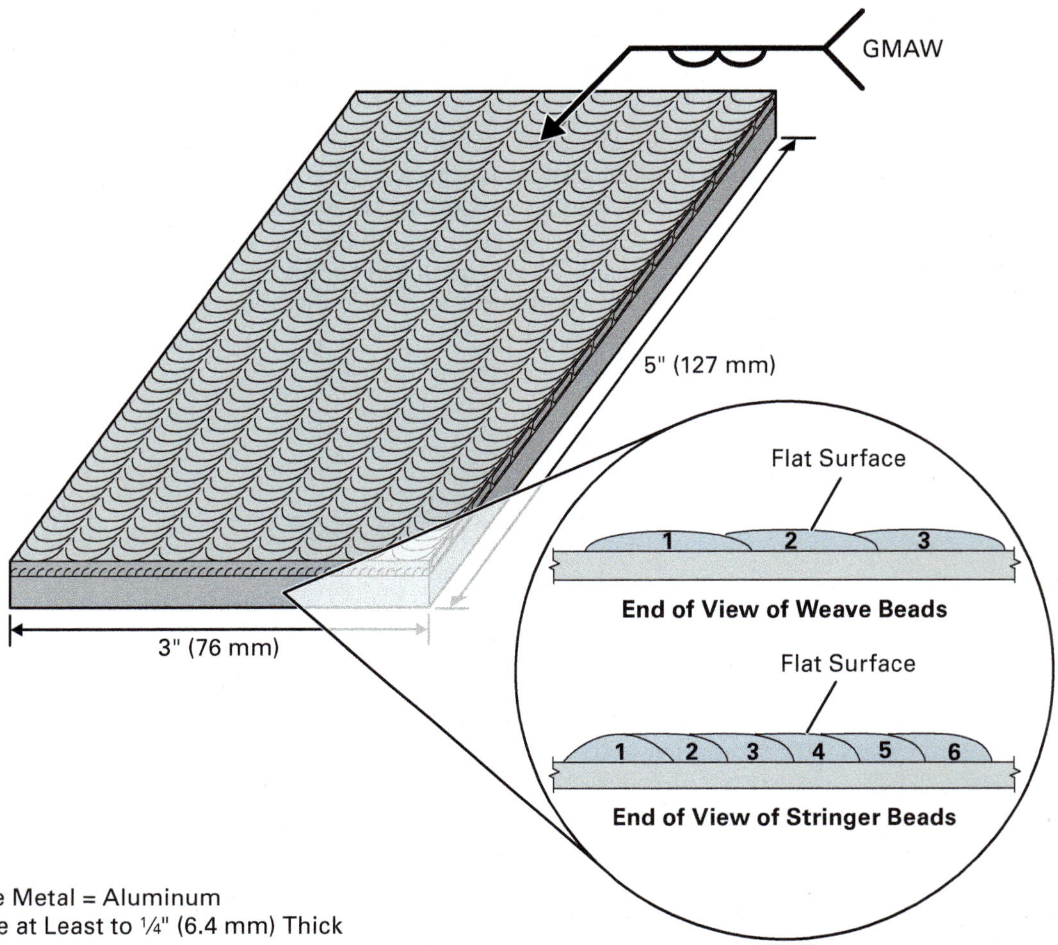

Note: Base Metal = Aluminum
Plate at Least to ¼" (6.4 mm) Thick

Criteria for Acceptance:

- Uniform, rippled appearance on the bead face __________
- Craters and restarts filled to the full cross section of the weld __________
- Uniform weld size ± 1⁄16" (1.6 mm) __________
- Smooth, flat transition with complete fusion at the toes of the weld __________
- Acceptable weld profiles in accordance with *Figure 21* of this module __________
- No excessive porosity with no pores larger than 3⁄32" (2.4 mm) __________
- No undercut greater than 1⁄32" (0.8 mm) deep or 10% of the base metal thickness, whichever is less __________
- No cracks __________

Figure A01

Performance Accreditation Tasks **Module 29401**

Make Multipass Fillet Welds on Aluminum Plate in the Flat (1F) Position

As directed by the instructor, use GMAW process with the appropriate aluminum wire to make a six-pass fillet weld using stringer beads on aluminum plate, as shown.

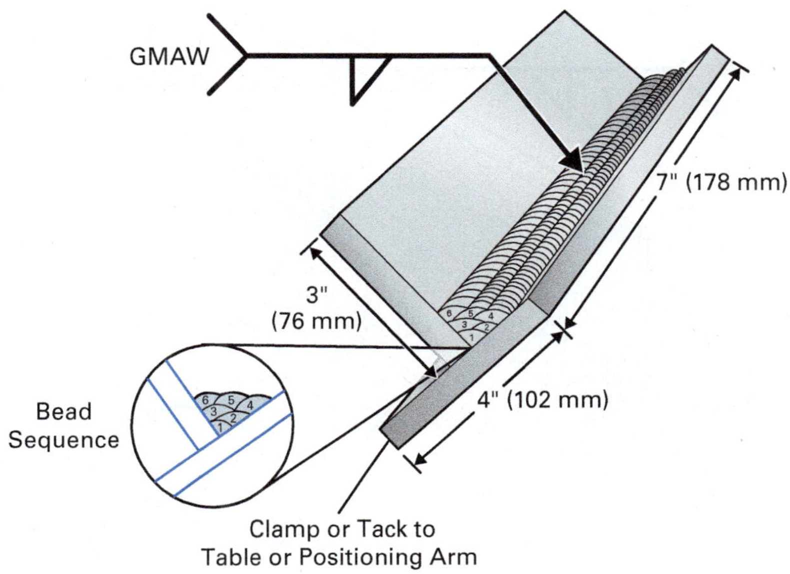

Criteria for Acceptance:

- Uniform, rippled appearance on the bead face ________
- Craters and restarts filled to the full cross section of the weld ________
- Uniform weld size ± 1⁄16" (1.6 mm) ________
- Smooth, flat transition with complete fusion at the toes of the weld ________
- Acceptable weld profiles in accordance with *Figure 21* of this module ________
- No excessive porosity with no pores larger than 3⁄32" (2.4 mm) ________
- No undercut greater than 1⁄32" (0.8 mm) deep or 10% of the base metal thickness, whichever is less ________
- No overlap ________

Figure A02

Performance Accreditation Tasks

Module 29401

Make Multipass Fillet Welds on Aluminum Plate in the Horizontal (2F) Position

As directed by the instructor, use GMAW process with the appropriate aluminum wire to make a six-pass fillet weld using stringer beads on aluminum plate, as shown.

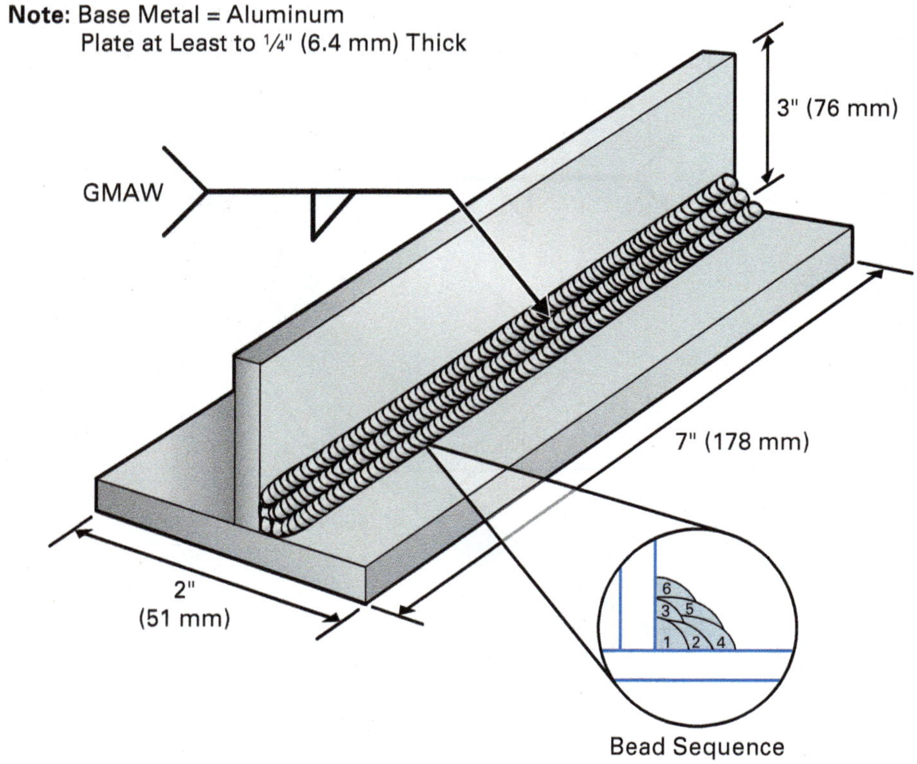

Criteria for Acceptance:

- Uniform, rippled appearance on the bead face __________
- Craters and restarts filled to the full cross section of the weld __________
- Uniform weld size ± 1⁄16" (1.6 mm) __________
- Smooth, flat transition with complete fusion at the toes of the weld __________
- Acceptable weld profiles in accordance with *Figure 21* of this module __________
- No excessive porosity with no pores larger than 3⁄32" (2.4 mm) __________
- No undercut greater than 1⁄32" (0.8 mm) deep or 10% of the base metal thickness, whichever is less __________
- No overlap __________

Figure A03

Performance Accreditation Tasks **Module 29401**

Make Multipass Fillet Welds on Aluminum Plate in the Vertical (3F) Position

As directed by the instructor, use GMAW process with the appropriate aluminum wire to make a six-pass fillet weld using stringer beads on aluminum plate, as shown.

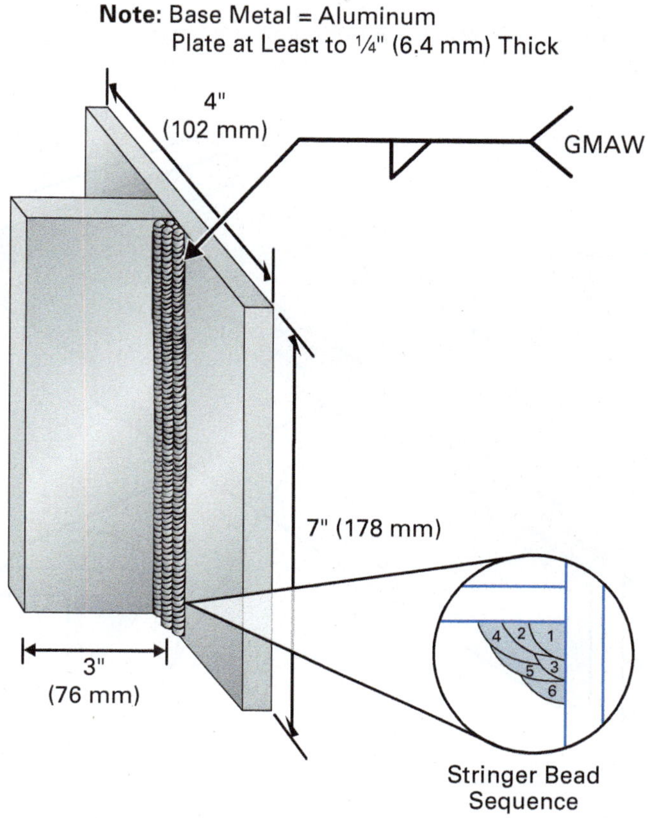

Criteria for Acceptance:

- Uniform, rippled appearance on the bead face ________
- Craters and restarts filled to the full cross section of the weld ________
- Uniform weld size ± 1⁄16" (1.6 mm) ________
- Smooth, flat transition with complete fusion at the toes of the weld ________
- Acceptable weld profiles in accordance with *Figure 21* of this module ________
- No excessive porosity with no pores larger than 3⁄32" (2.4 mm) ________
- No undercut greater than 1⁄32" (0.8 mm) deep or 10% of the base metal thickness, whichever is less ________
- No cracks ________

Figure A04

Performance Accreditation Tasks

Module 29401

Make Multipass Fillet Welds on Aluminum Plate in the Overhead (4F) Position

As directed by the instructor, use GMAW process with the appropriate aluminum wire to make a six-pass fillet weld using stringer beads on aluminum plate, as shown.

Note: Base Metal = Aluminum
Plate at Least to ¼" (6.4 mm) Thick

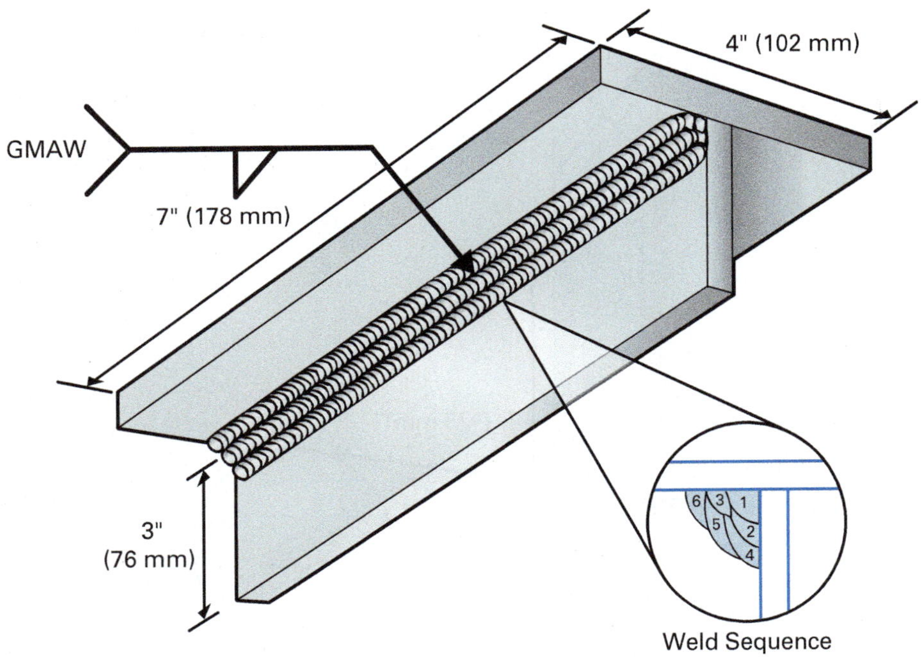

Criteria for Acceptance:

- Uniform, rippled appearance on the bead face ________
- Craters and restarts filled to the full cross section of the weld ________
- Uniform weld size ± 1⁄16" (1.6 mm) ________
- Smooth, flat transition with complete fusion at the toes of the weld ________
- Acceptable weld profiles in accordance with *Figure 21* of this module ________
- No excessive porosity with no pores larger than 3⁄32" (2.4 mm) ________
- No undercut greater than 1⁄32" (0.8 mm) deep or 10% of the base metal thickness, whichever is less ________
- No cracks ________

Figure A05

Performance Accreditation Tasks **Module 29401**

Make Multipass V-Groove Welds on Aluminum Plate with Backing in the Flat (1F) Position

As directed by the instructor, use GMAW process with the appropriate aluminum wire to make a multipass groove weld using stringer beads on aluminum plate with backing, as shown.

Note: Base Metal, Aluminum Plate
¼" to ½" (6.4 mm to 12.7 mm) Thick

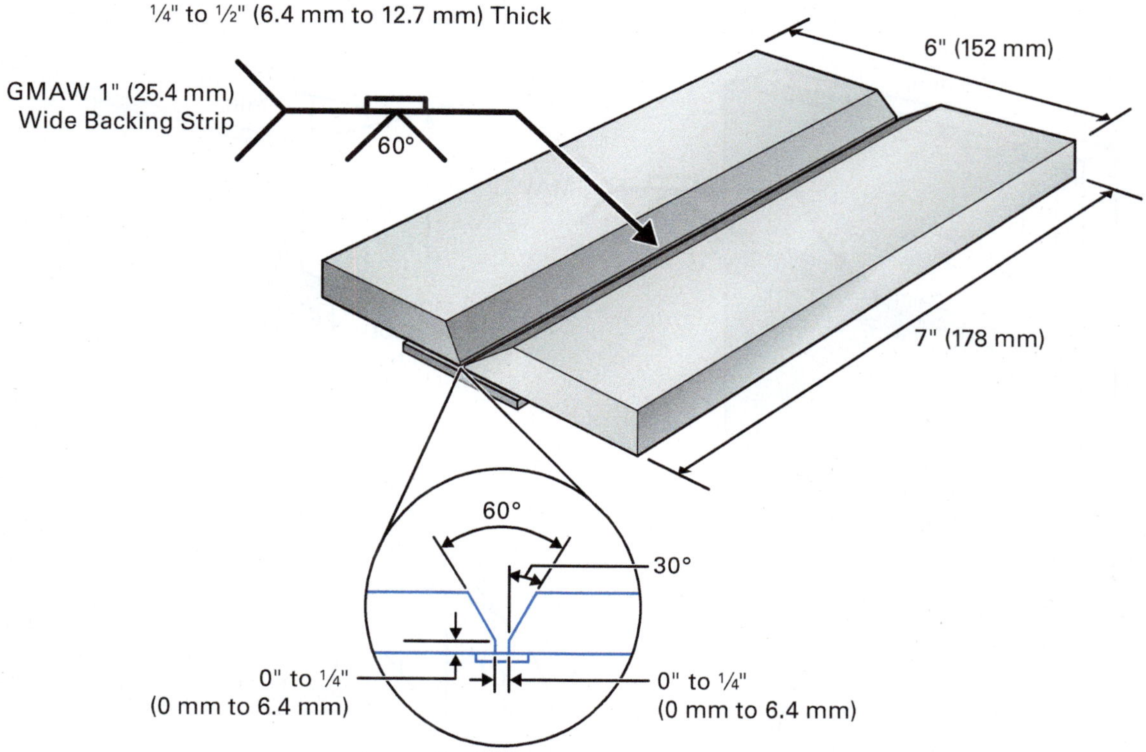

Joint Details

Criteria for Acceptance:

- Uniform, rippled appearance on the bead face ______
- Craters and restarts filled to the full cross section of the weld ______
- Uniform weld size ± 1⁄16" (1.6 mm) ______
- Smooth, flat transition with complete fusion at the toes of the weld ______
- Acceptable weld profiles in accordance with *Figure 28* of this module ______
- No excessive porosity with no pores larger than 3⁄32" (2.4 mm) ______
- No undercut greater than 1⁄32" (0.8 mm) deep or 10% of the base metal thickness, whichever is less ______
- No overlap ______
- No cracks ______

Figure A06

Performance Accreditation Tasks

Module 29401

Make Multipass V-Groove Welds on Aluminum Plate with Backing in the Horizontal (2G) Position

As directed by the instructor, use GMAW process with the appropriate aluminum wire to make a multipass groove weld using stringer beads on aluminum plate with backing, as shown.

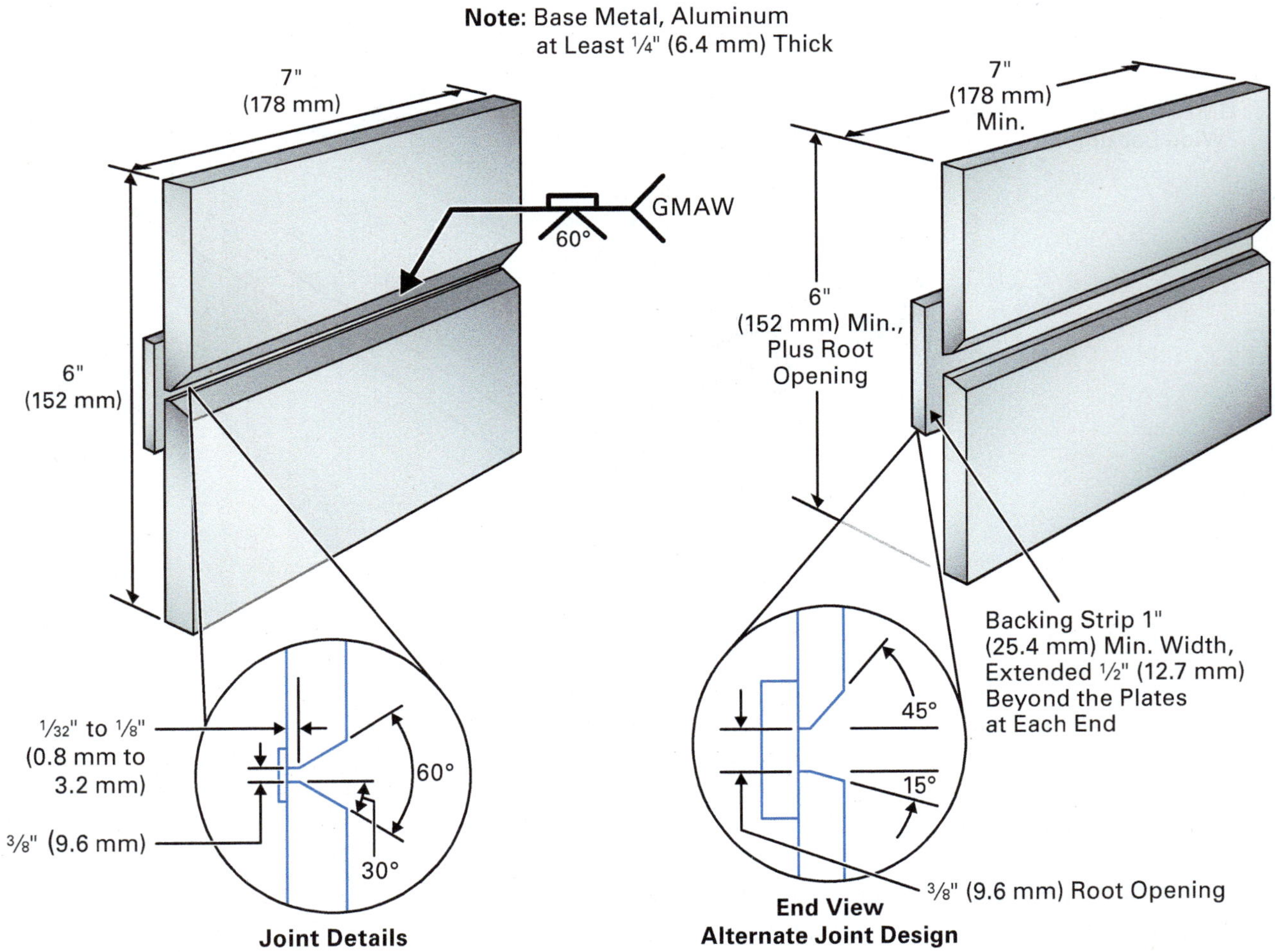

Criteria for Acceptance:

- Uniform, rippled appearance on the bead face ________
- Craters and restarts filled to the full cross section of the weld ________
- Uniform weld size ± 1⁄16" (1.6 mm) ________
- Smooth, flat transition with complete fusion at the toes of the weld ________
- Acceptable weld profiles in accordance with *Figure 28* of this module ________
- No excessive porosity with no pores larger than 3⁄32" (2.4 mm) ________
- No undercut greater than 1⁄32" (0.8 mm) deep or 10% of the base metal thickness, whichever is less ________
- No overlap ________
- No cracks ________

Figure A07

Performance Accreditation Tasks

Module 29401

Make Multipass V-Groove Welds on Aluminum Plate with Backing in the Vertical (3G) Position

As directed by the instructor, use GMAW process with the appropriate aluminum wire to make a multipass groove weld using stringer beads on aluminum plate with a 1" (25 mm) backing, as shown.

Note: Run the root vertical up.

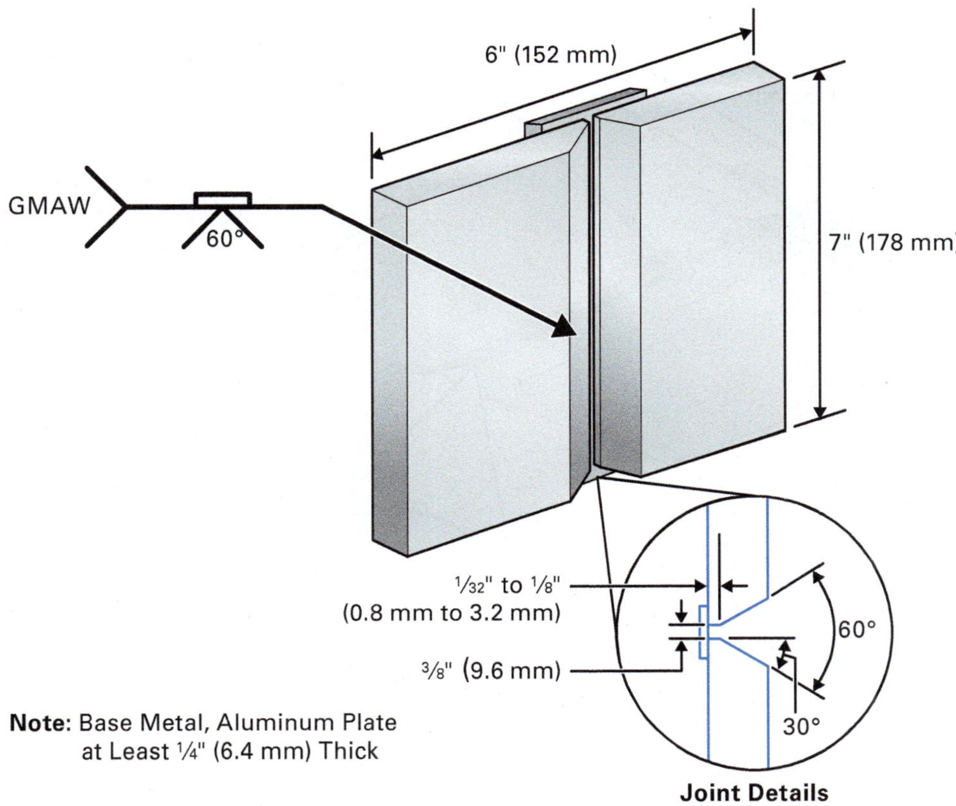

Criteria for Acceptance:

- Uniform, rippled appearance on the bead face ________
- Craters and restarts filled to the full cross section of the weld ________
- Uniform weld size ± 1⁄16" (1.6 mm) ________
- Smooth, flat transition with complete fusion at the toes of the weld ________
- Acceptable weld profiles in accordance with *Figure 28* of this module ________
- No excessive porosity with no pores larger than 3⁄32" (2.4 mm) ________
- No undercut greater than 1⁄32" (0.8 mm) deep or 10% of the base metal thickness, whichever is less ________
- No overlap ________
- No cracks ________

Figure A08

Performance Accreditation Tasks

Module 29401

Make Multipass V-Groove Welds on Aluminum Plate in the Backing in the Overhead (4F) Position

As directed by the instructor, use GMAW process with the appropriate aluminum wire to make a multipass groove weld using stringer beads on aluminum plate with a 1" (25 mm) backing, as shown.

Note: Base Metal, Aluminum Plate at Least ¼" (6.4 mm) Thick

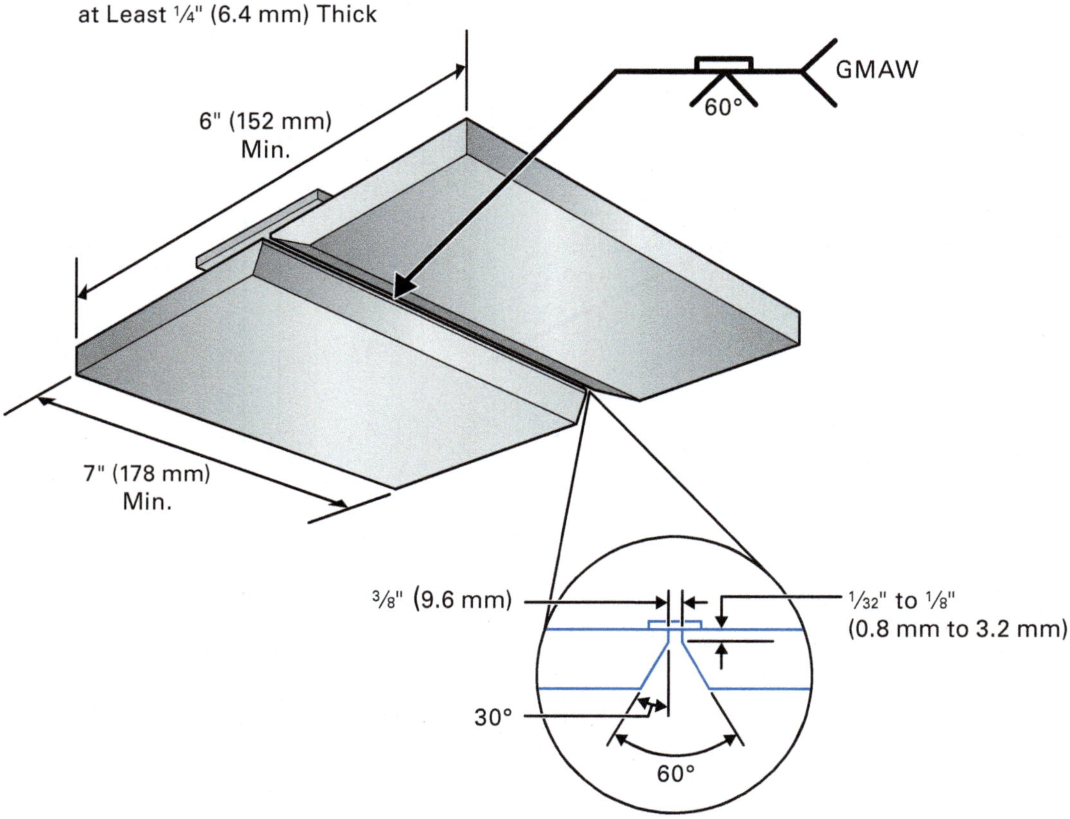

Joint Details

Criteria for Acceptance:

- Uniform, rippled appearance on the bead face ______
- Craters and restarts filled to the full cross section of the weld ______
- Uniform weld size ± 1⁄16" (1.6 mm) ______
- Smooth, flat transition with complete fusion at the toes of the weld ______
- Acceptable weld profiles in accordance with *Figure 28* of this module ______
- No excessive porosity with no pores larger than 3⁄32" (2.4 mm) ______
- No undercut greater than 1⁄32" (0.8 mm) deep or 10% of the base metal thickness, whichever is less ______
- No overlap ______
- No cracks ______

Figure A09

APPENDIX 29404A

Performance Accreditation Tasks

Performance Accreditation Tasks (PATs) provide specific acceptable criteria for performance and help to ensure a true competency-based welding program for students.

PATs 1 and 2 have no AWS correlation. PAT 3 corresponds to and supports learning objectives in the following:

- *AWS QC10, Specifications for Qualification and Certification of SENSE Level I—Entry Welders*
- *AWS QC11, Specifications for Qualification and Certification of SENSE Level II—Advanced Welders*

Each of the tasks are designed to evaluate your ability to run stringer beads, weave beads, and overlapping beads. In addition, fillet and V-groove welds must be completed with GMAW equipment in four standard test positions using aluminum filler wire of the appropriate diameter and shielding gas. Perform each task when you are instructed to do so by your instructor. As you complete each task, show it to your instructor for evaluation. Do not proceed to the next task until told to do so by your instructor.

Performance Accreditation Tasks **Module 29404**

V-Groove Welds on Aluminum Pipe with Backing in the 2G Position

Using aluminum filler wire of the appropriate diameter, proper shielding gas, and stringer beads, make V-groove welds on aluminum pipe in the 2G position.

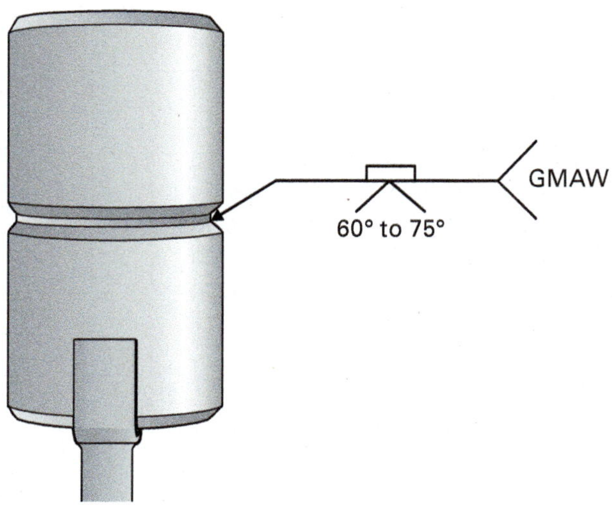

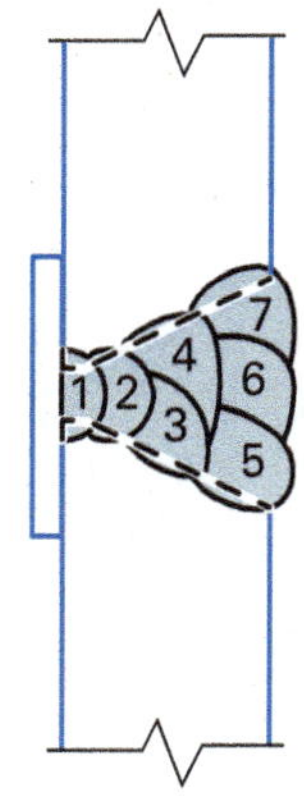

Stringer Bead Sequence

Criteria for Acceptance:

- Uniform, rippled appearance on the bead face __________
- Craters and restarts filled to the full cross section of the weld __________
- Acceptable weld profiles in accordance with the applicable codes and standards (Refer to *Figure 11* for examples of unacceptable pipe groove weld profiles) __________
- Smooth transition with complete fusion at the toes of the weld __________
- No excessive porosity with no pores larger than 3⁄32" (2.4 mm) __________
- No undercut greater than 1⁄32" (0.8 mm) deep or 10% of the base metal thickness, whichever is less __________
- No underfill __________
- No cracks __________
- No overlap __________
- No incomplete fusion __________

Figure A01

Performance Accreditation Tasks

Module 29404

V-Groove Welds on Aluminum Pipe with Backing in the 5G Position

Using aluminum filler wire of the appropriate diameter, proper shielding gas, and stringer weave beads, make V-groove welds on aluminum pipe in the 5G position.

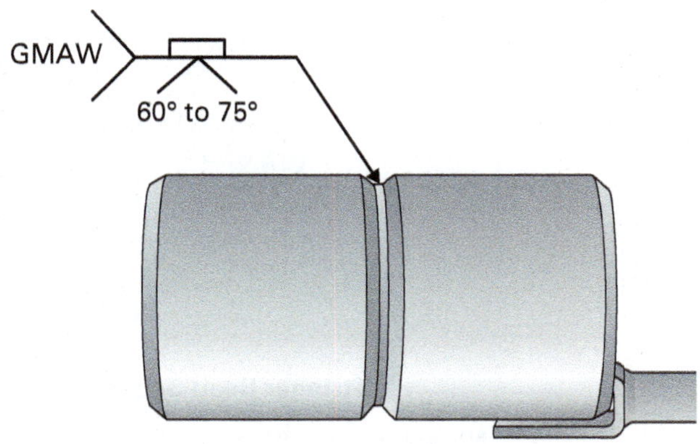

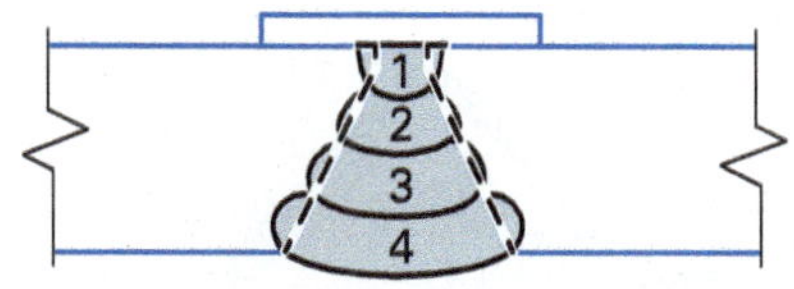

Weave Bead Sequence

Stringer Bead Sequence

Criteria for Acceptance:

- Uniform, rippled appearance on the bead face __________
- Craters and restarts filled to the full cross section of the weld __________
- Acceptable weld profiles in accordance with the applicable codes and standards (Refer to *Figure 11* for examples of unacceptable pipe groove weld profiles) __________
- Smooth transition with complete fusion at the toes of the weld __________
- No excessive porosity with no pores larger than 3⁄32" (2.4 mm) __________
- No undercut greater than 1⁄32" (0.8 mm) deep or 10% of the base metal thickness, whichever is less __________
- No underfill __________
- No cracks __________
- No overlap __________
- No incomplete fusion __________

Figure A02

Performance Accreditation Tasks

Module 29404

V-Groove Welds on Aluminum Pipe with Backing in the 6G Position

Using aluminum filler wire of the appropriate diameter, proper shielding gas, and stringer weave beads, make V-groove welds on aluminum pipe in the 6G position.

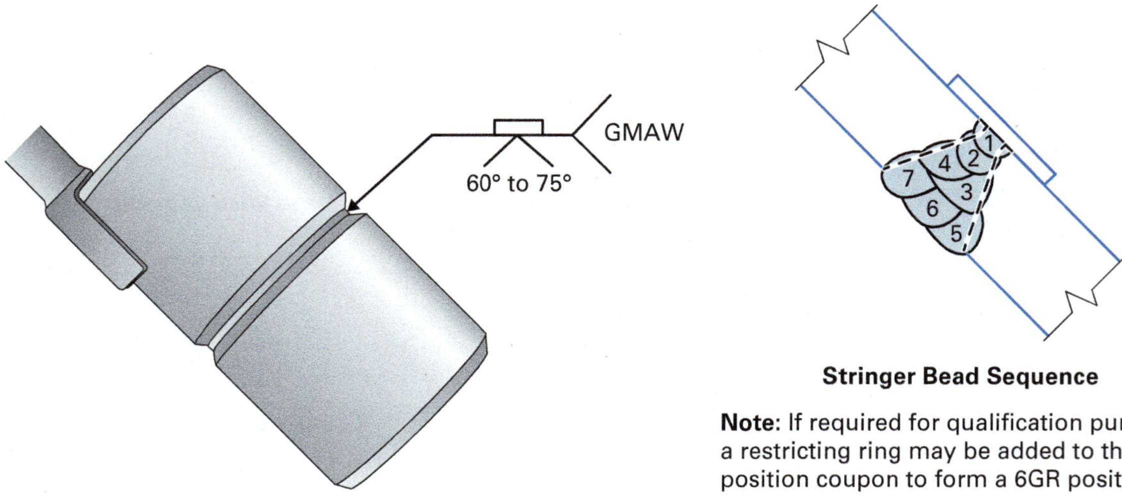

Stringer Bead Sequence

Note: If required for qualification purposes, a restricting ring may be added to the 6G position coupon to form a 6GR position coupon.

Criteria for Acceptance:

- Uniform, rippled appearance on the bead face ____________
- Craters and restarts filled to the full cross section of the weld ____________
- Acceptable weld profiles in accordance with the applicable codes and standards (Refer to *Figure 11* for examples of unacceptable pipe groove weld profiles) ____________
- Smooth transition with complete fusion at the toes of the weld ____________
- No excessive porosity with no pores larger than 3⁄32" (2.4 mm) ____________
- No undercut greater than 1⁄32" (0.8 mm) deep or 10% of the base metal thickness, whichever is less ____________
- No underfill ____________
- No cracks ____________
- No overlap ____________
- No incomplete fusion ____________

Figure A03

APPENDIX 29402A

Performance Accreditation Tasks

Performance Accreditation Tasks (PATs) provide specific acceptable criteria for performance and help to ensure a true competency-based welding program for students.

PAT 7 has no AWS correlation. PATs 1, 2, 3, and 6 correspond to and support learning objectives in *AWS EG2.0, Guide for the Training of Welding Personnel: SENSE Level I—Entry Welders*. PATs 3, 4, 8, and 9 correspond to and support learning objectives *AWS EG3.0, Guide for the Training of Welding Personnel: SENSE Level II—Advanced Welders*. PATs 1 and 8 also correspond to *AWS QC10, Specification for Qualification and Certification of SENSE Level I—Entry Welders*. Additionally, PAT 8 corresponds to *AWS QC11, Specification for Qualification and Certification of SENSE Level II—Advanced Welders*.

Each of the tasks are designed to evaluate your ability to run stringer beads, weave beads, and overlapping beads. In addition, fillet and V-groove welds must be completed with GMAW equipment in four standard test positions using aluminum filler wire of the appropriate diameter and shielding gas. Perform each task when you are instructed to do so by your instructor. As you complete each task, show it to your instructor for evaluation. Do not proceed to the next task until told to do so by your instructor.

Performance Accreditation Tasks

Module 29402

Weld A Pad On Aluminum Plate in the Flat (1G) Position Using GTAW Stringer Beads

As directed by the instructor, use the GTAW process with the appropriate aluminum filler wire to make the following welds on aluminum plate: stringer beads, weave beads, weld restarts, weld terminations, and overlapping beads.

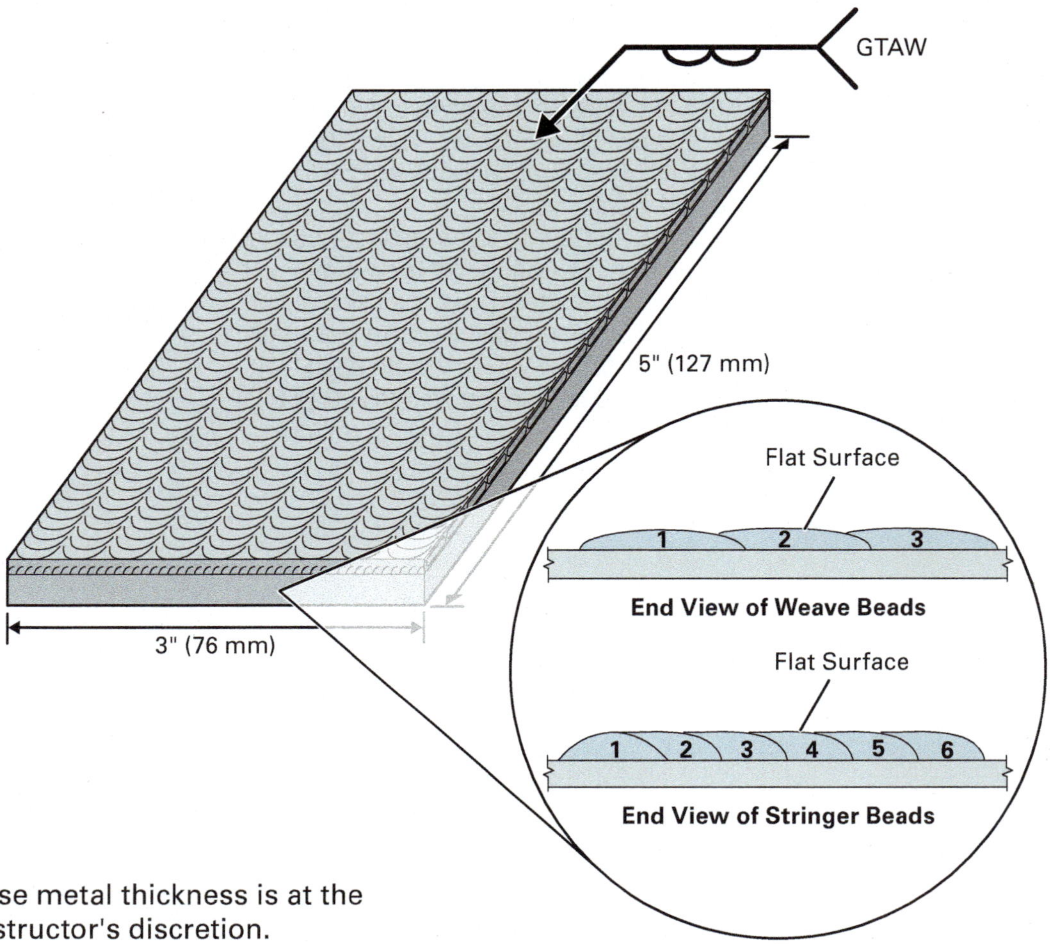

Note: Base metal thickness is at the instructor's discretion.

Criteria for Acceptance:

- Uniform, rippled appearance on the bead face ____________
- Craters and restarts filled to the full cross section of the weld ____________
- Uniform weld size ±1⁄16" (1.6 mm) ____________
- Smooth, flat transition with complete fusion at the toes of the weld ____________
- Acceptable weld profiles for padding in accordance with *Figure 16* of this module ____________
- No excessive porosity with no pores larger than 3⁄32" (2.4 mm) ____________
- No undercut greater than 1⁄32" (0.8 mm) deep or 10% of the base metal thickness, whichever is less ____________
- No cracks ____________

Figure A01

Performance Accreditation Tasks

Module 29402

Make Multipass Fillet Welds on Aluminum Plate in the Flat (1F) Position

As directed by the instructor, use the GTAW process with the appropriate aluminum wire to make a six-pass fillet weld using stringer beads on aluminum plate, as shown.

Note: Base metal thickness is at the instructor's discretion.

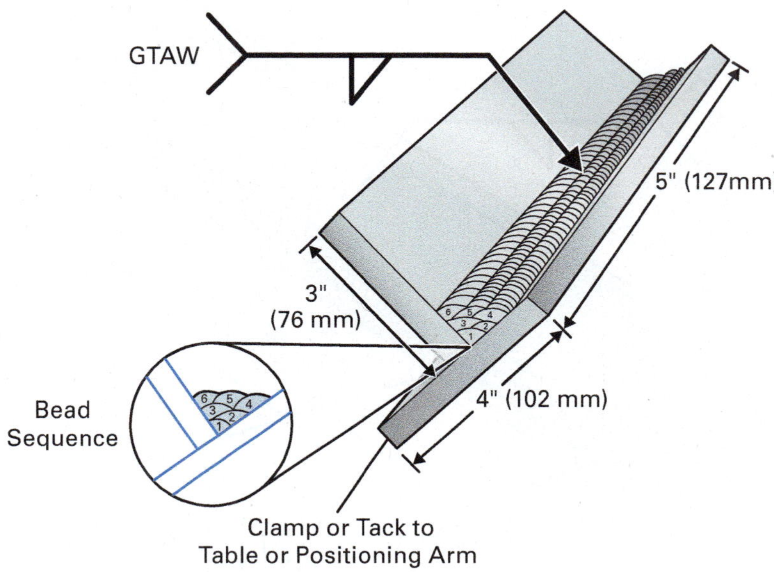

Criteria for Acceptance:

- Uniform, rippled appearance on the bead face ______________
- Craters and restarts filled to the full cross section of the weld ______________
- Uniform weld size ±1⁄16" (1.6 mm) ______________
- Smooth, flat transition with complete fusion at the toes of the weld ______________
- Acceptable weld profiles in accordance with *Figure 19* of this module ______________
- No excessive porosity with no pores larger than 3⁄32" (2.4 mm) ______________
- No undercut greater than 1⁄32" (0.8 mm) deep or 10% of the base metal thickness, whichever is less ______________
- No overlap ______________

Figure A02

Performance Accreditation Tasks

Module 29402

Make Multipass Fillet Welds on Aluminum Plate in the Horizontal (2F) Position

As directed by the instructor, use the GTAW process with the appropriate aluminum wire to make a six-pass fillet weld using stringer beads on aluminum plate, as shown.

Note: Base metal thickness is at the instructor's discretion.

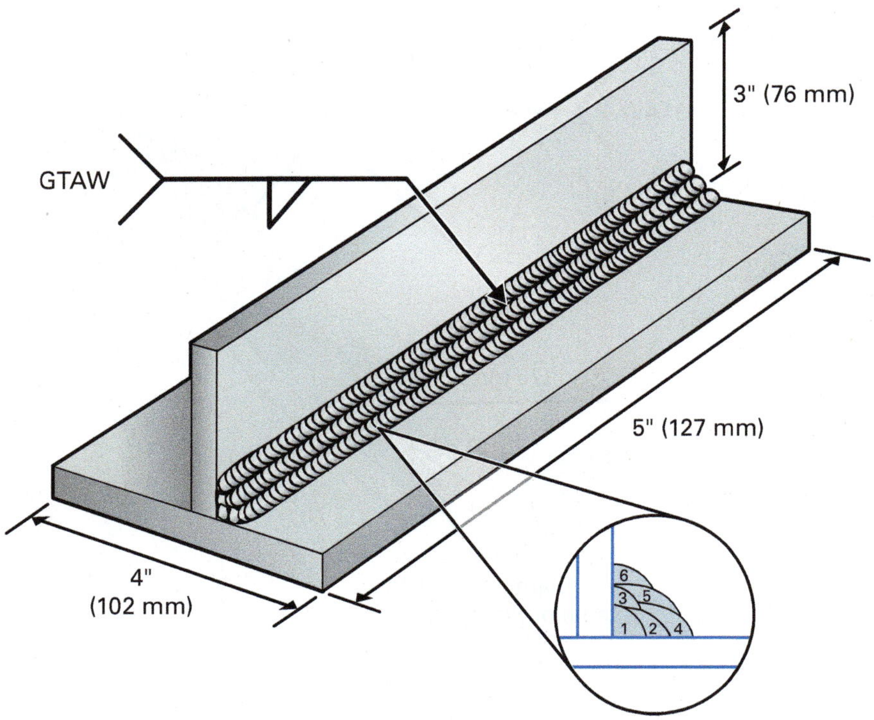

Criteria for Acceptance:

- Uniform, rippled appearance on the bead face ____________
- Craters and restarts filled to the full cross section of the weld ____________
- Uniform weld size ±1⁄16" (1.6 mm) ____________
- Smooth, flat transition with complete fusion at the toes of the weld ____________
- Acceptable weld profiles in accordance with *Figure 19* of this module ____________
- No excessive porosity with no pores larger than 3⁄32" (2.4 mm) ____________
- No undercut greater than 1⁄32" (0.8 mm) deep or 10% of the base metal thickness, whichever is less ____________
- No overlap ____________

Figure A03

Performance Accreditation Tasks

Module 29402

Make Multipass Fillet Welds on Aluminum Plate in the Vertical (3F) Position

As directed by the instructor, use the GTAW process with the appropriate aluminum wire to make a six-pass fillet weld using stringer beads on aluminum plate, as shown.

Note: Base metal thickness is at the instructor's discretion.

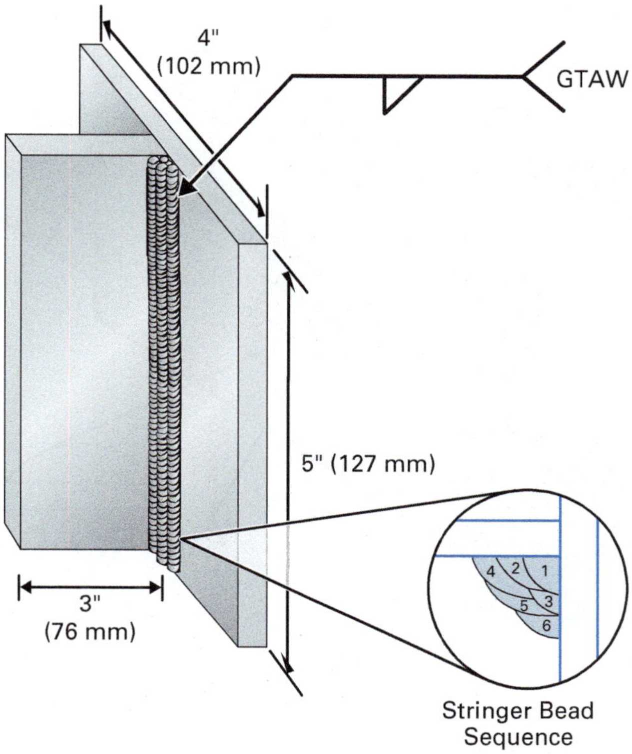

Criteria for Acceptance:

- Uniform, rippled appearance on the bead face ______
- Craters and restarts filled to the full cross section of the weld ______
- Uniform weld size ±1⁄16" (1.6 mm) ______
- Smooth, flat transition with complete fusion at the toes of the weld ______
- Acceptable weld profiles in accordance with *Figure 19* of this module ______
- No excessive porosity with no pores larger than 3⁄32" (2.4 mm) ______
- No undercut greater than 1⁄32" (0.8 mm) deep or 10% of the base metal thickness, whichever is less ______
- No cracks ______

Figure A04

Performance Accreditation Tasks

Module 29402

Make Multipass Fillet Welds on Aluminum Plate in the Overhead (4F) Position

As directed by the instructor, use the GTAW process with the appropriate aluminum wire to make a six-pass fillet weld using stringer beads on aluminum plate, as shown.

Note: Base metal thickness is at the instructor's discretion.

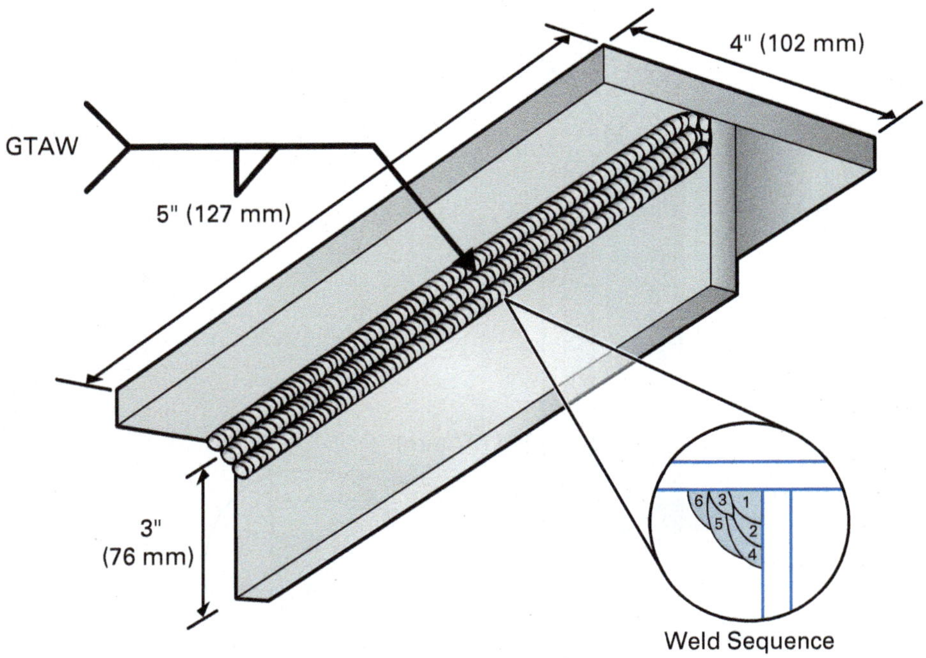

Criteria for Acceptance:

- Uniform, rippled appearance on the bead face ________
- Craters and restarts filled to the full cross section of the weld ________
- Uniform weld size ±1⁄16" (1.6 mm) ________
- Smooth, flat transition with complete fusion at the toes of the weld ________
- Acceptable weld profiles in accordance with *Figure 19* of this module ________
- No excessive porosity with no pores larger than 3⁄32" (2.4 mm) ________
- No undercut greater than 1⁄32" (0.8 mm) deep or 10% of the base metal thickness, whichever is less ________
- No cracks ________

Figure A05

Performance Accreditation Tasks

Module 29402

Make Multipass V-Groove Welds on Aluminum Plate with Backing in the Flat (1F) Position

As directed by the instructor, use the GTAW process with the appropriate aluminum wire to make a multipass groove weld using stringer beads on aluminum plate with backing, as shown.

Note: Base metal thickness is at the instructor's discretion.

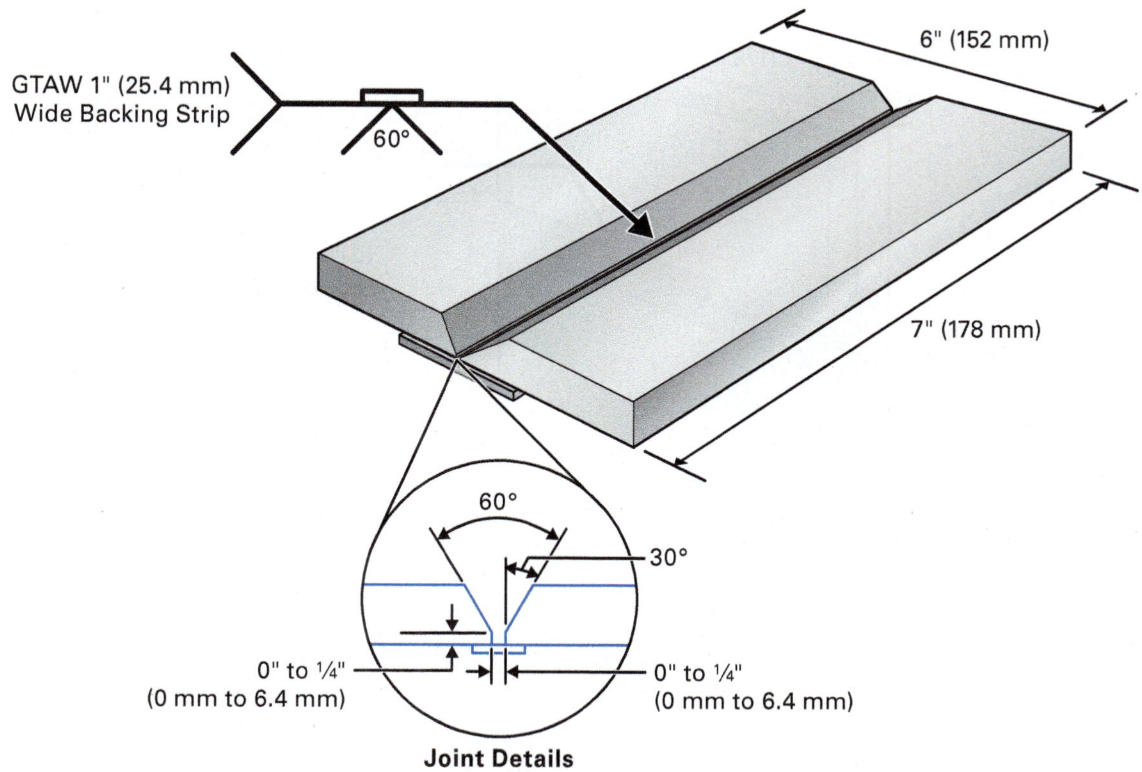

Criteria for Acceptance:

- Uniform, rippled appearance on the bead face ____________
- Craters and restarts filled to the full cross section of the weld ____________
- Uniform weld size ±1⁄16" (1.6 mm) ____________
- Smooth, flat transition with complete fusion at the toes of the weld ____________
- Acceptable weld profile in accordance with *Figure 25* of this module ____________
- No excessive porosity with no pores larger than 3⁄32" (2.4 mm) ____________
- No undercut greater than 1⁄32" (0.8 mm) deep or 10% of the base metal thickness, whichever is less ____________
- No overlap ____________
- No cracks ____________

Figure A06

Performance Accreditation Tasks

Module 29402

Make Multipass V-Groove Welds on Aluminum Plate with Backing in the Horizontal (2G) Position

As directed by the instructor, use the GTAW process with the appropriate aluminum wire to make a multipass groove weld using stringer beads on aluminum plate with backing, as shown.

Note: Base metal thickness is at the instructor's discretion.

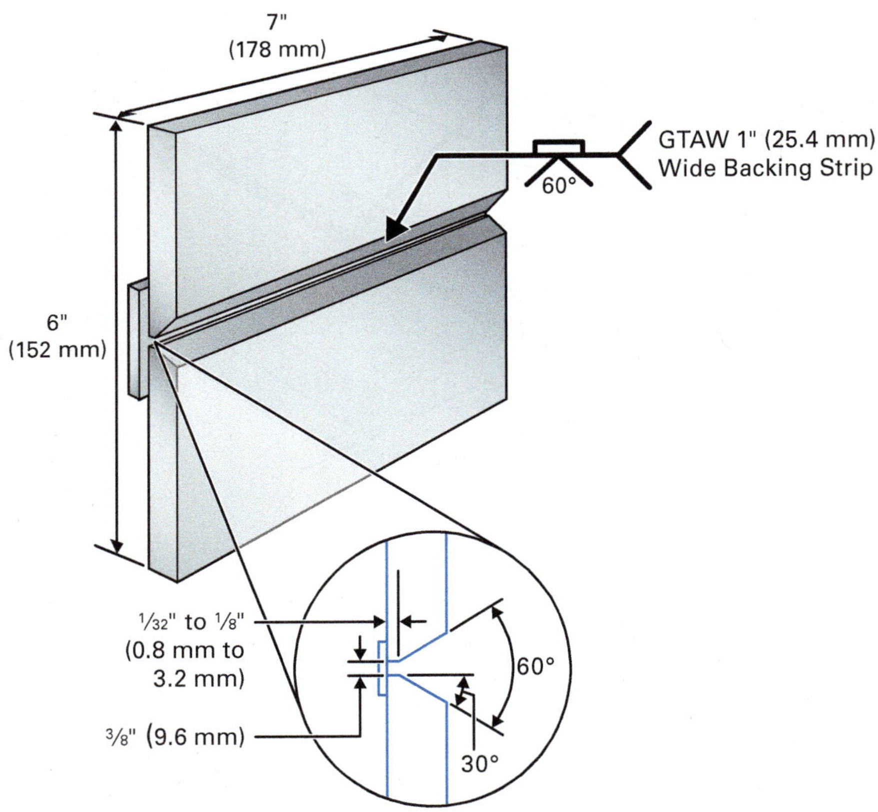

Joint Details

Criteria for Acceptance:

- Uniform, rippled appearance on the bead face ______
- Craters and restarts filled to the full cross section of the weld ______
- Uniform weld size ±1⁄16" (1.6 mm) ______
- Smooth, flat transition with complete fusion at the toes of the weld ______
- Acceptable weld profile in accordance with *Figure 25* of this module ______
- No excessive porosity with no pores larger than 3⁄32" (2.4 mm) ______
- No undercut greater than 1⁄32" (0.8 mm) deep or 10% of the base metal thickness, whichever is less ______
- No overlap ______
- No cracks ______

Figure A07

Performance Accreditation Tasks **Module 29402**

Make Multipass V-Groove Welds on Aluminum Plate with Backing in the Vertical (3G) Position

As directed by the instructor, use the GTAW process with the appropriate aluminum wire to make a multipass groove weld on aluminum plate with backing, as shown.

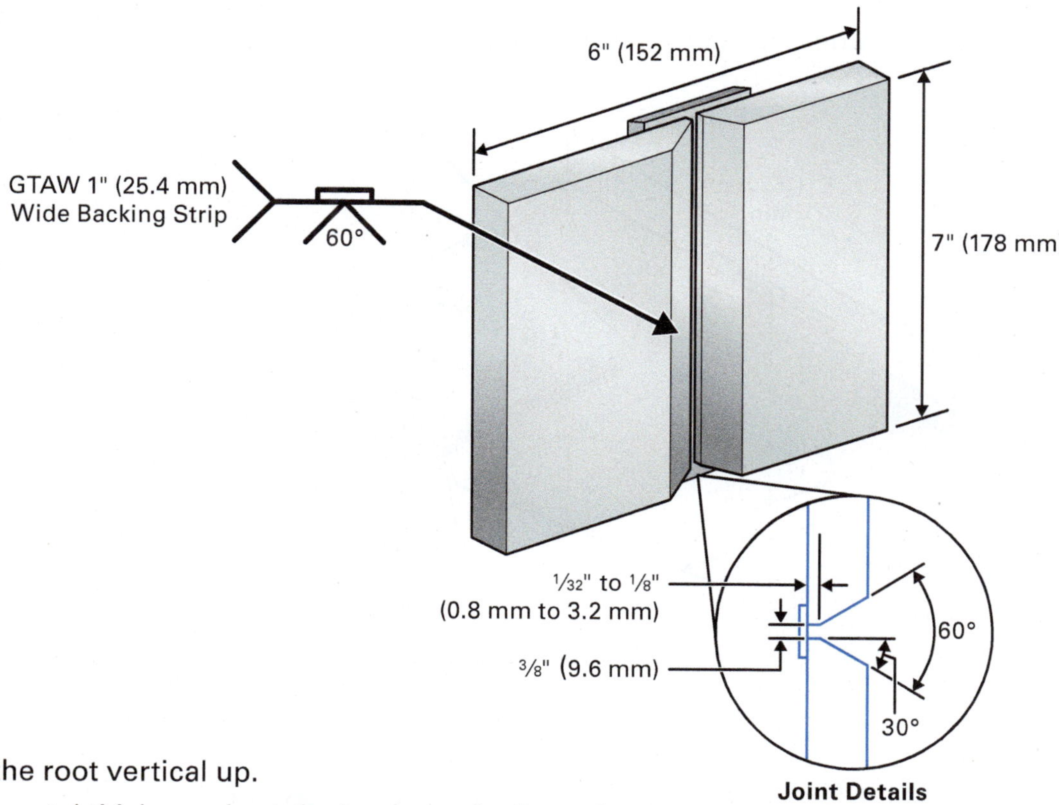

Notes:

- Run the root vertical up.
- Base metal thickness is at the instructor's discretion.

Criteria for Acceptance:

- Uniform, rippled appearance on the bead face ________
- Craters and restarts filled to the full cross section of the weld ________
- Uniform weld size ±1⁄16" (1.6 mm) ________
- Smooth, flat transition with complete fusion at the toes of the weld ________
- Acceptable weld profile in accordance with *Figure 25* of this module ________
- No excessive porosity with no pores larger than 3⁄32" (2.4 mm) ________
- No undercut greater than 1⁄32" (0.8 mm) deep or 10% of the base metal thickness, whichever is less ________
- No overlap ________
- No cracks ________

Figure A08

Performance Accreditation Tasks

Module 29402

Make Multipass V-Groove Welds on Aluminum Plate with Backing in the Overhead (4F) Position

As directed by the instructor, use the GTAW process with the appropriate aluminum wire to make a multipass groove weld using stringer beads on aluminum plate with a 1" (25.4 mm) backing, as shown.

Note: Base metal thickness is at the instructor's discretion.

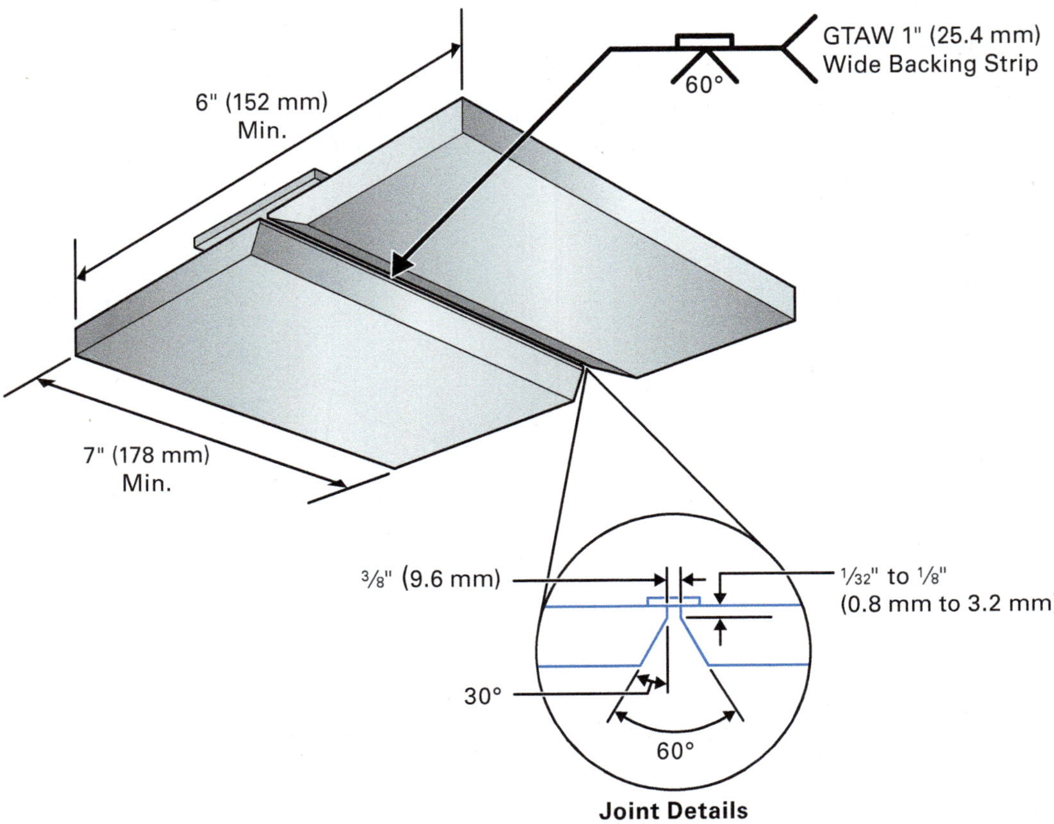

Criteria for Acceptance:

- Uniform, rippled appearance on the bead face ______
- Craters and restarts filled to the full cross section of the weld ______
- Uniform weld size ±1⁄16" (1.6 mm) ______
- Smooth, flat transition with complete fusion at the toes of the weld ______
- Acceptable weld profile in accordance with *Figure 25* of this module ______
- No excessive porosity with no pores larger than 3⁄32" (2.4 mm) ______
- No undercut greater than 1⁄32" (0.8 mm) deep or 10% of the base metal thickness, whichever is less ______
- No overlap ______
- No cracks ______

Figure A09

APPENDIX 29403A

Performance Accreditation Tasks

Performance Accreditation Tasks (PATs) provide specific acceptable criteria for performance and help to ensure a true competency-based welding program for students.

PATs 1 and 2 have no AWS correlation. PAT 3 corresponds to and supports learning objectives in *AWS QC11, Specifications for Qualification and Certification of SENSE Level II—Advanced Welders*.

Each of the tasks are designed to evaluate your ability to run stringer beads, weave beads, and overlapping beads. In addition, fillet and V-groove welds must be completed with GMAW equipment in four standard test positions using aluminum filler wire of the appropriate diameter and shielding gas. Perform each task when you are instructed to do so by your instructor. As you complete each task, show it to your instructor for evaluation. Do not proceed to the next task until told to do so by your instructor.

Performance Accreditation Tasks **Module 29403**

U-Groove Welds on Aluminum Pipe in the 2G Position

As directed by the instructor, use the GTAW process with the appropriate aluminum filler wire, argon shielding gas, and stringer beads to make a multipass groove weld on aluminum pipe in the 2G position.

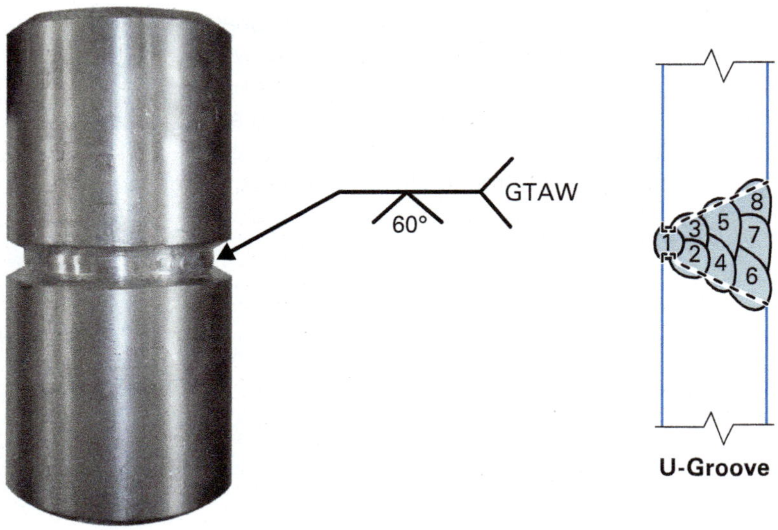

(A) (B)

Criteria for Acceptance:

- Uniform appearance on the bead face ____________
- Craters and restarts filled to the full cross section of the weld ____________
- Acceptable weld profile in accordance with *Figure 15* of this module ____________
- Maximum of 1⁄8" (3.2 mm) face and root reinforcement ____________
- Smooth transition with complete fusion at the toes of the weld ____________
- No porosity to exceed 3⁄32" (2.4 mm) ____________
- Undercut 1⁄32" (0.8 mm) or less than 10% of base metal thickness ____________
- No cracks ____________
- No overlap ____________
- No incomplete fusion ____________

Figure A01
Source: Terry Lowe (A)

Performance Accreditation Tasks **Module 29403**

U-Groove Welds on Aluminum Pipe in the 5G Position

As directed by the instructor, use the GTAW process with the appropriate aluminum filler wire, argon shielding gas, and stringer beads to make a multipass groove weld on aluminum pipe in the 5G position.

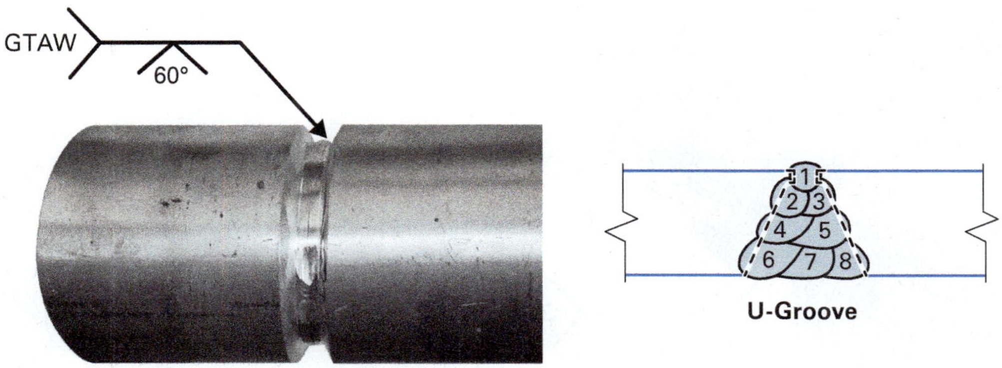

(A) **(B)**

Criteria for Acceptance:

- Uniform appearance on the bead face ____________
- Craters and restarts filled to the full cross section of the weld ____________
- Acceptable weld profile in accordance with *Figure 15* of this module ____________
- Maximum of $\frac{1}{8}$" (3.2 mm) face and root reinforcement ____________
- Smooth transition with complete fusion at the toes of the weld ____________
- No porosity to exceed $\frac{3}{32}$" (2.4 mm) ____________
- Undercut $\frac{1}{32}$" (0.8 mm) or less than 10% of base metal thickness ____________
- No cracks ____________
- No overlap ____________
- No incomplete fusion ____________

Figure A02
Source: Terry Lowe (A)

Performance Accreditation Tasks

Module 29403

U-Groove Welds on Aluminum Pipe in the 6G (or 6GR) Position

As directed by the instructor, use the GTAW process with the appropriate aluminum filler wire, argon shielding gas, and stringer beads to make a multipass groove weld on aluminum pipe in the 6G (or 6GR) position.

Note: If required for qualification purposes, a restricting ring may be added to the 6G position coupon to form a 6GR position coupon.

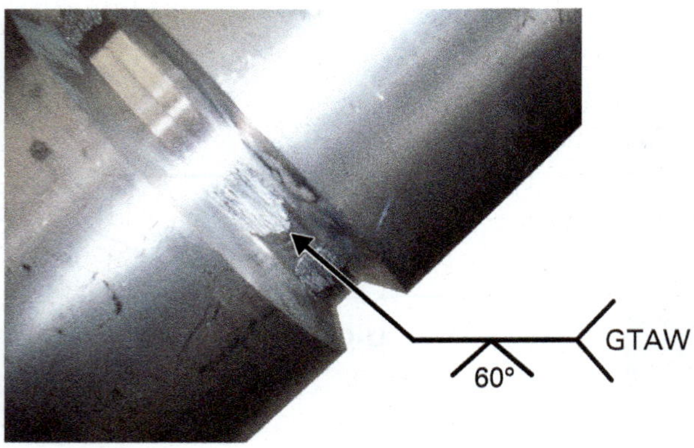

(A)

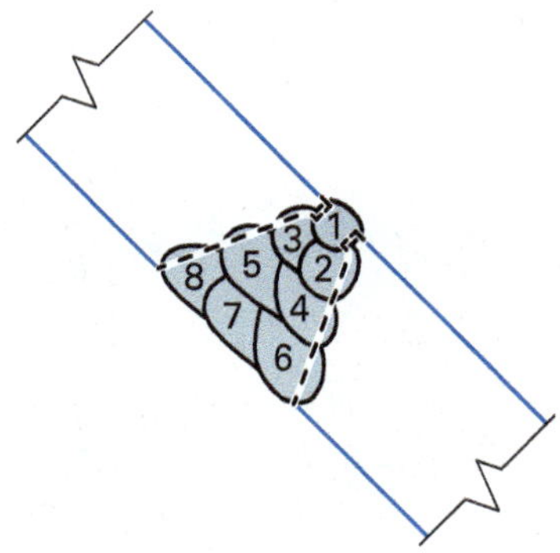

U-Groove

(B)

Criteria for Acceptance:

- Uniform appearance on the bead face ________
- Craters and restarts filled to the full cross section of the weld ________
- Acceptable weld profile in accordance with *Figure 15* of this module ________
- Maximum of 1⁄8" (3.2 mm) face and root reinforcement ________
- Smooth transition with complete fusion at the toes of the weld ________
- No porosity to exceed 3⁄32" (2.4 mm) ________
- Undercut 1⁄32" (0.8 mm) or less than 10% of base metal thickness ________
- No cracks ________
- No overlap ________
- No incomplete fusion ________

Figure A03
Source: Terry Lowe (A)

REVIEW QUESTION ANSWER KEYS

MODULE 29401

Answer	Section Reference
1. b	1.1.0
3. d	1.3.1
5. d	2.1.0
7. c	2.1.3
9. d	2.2.1
11. d	2.3.2
13. d	2.4.4
15. c	2.5.0

MODULE 29404

Answer	Section Reference
1. a	1.0.0
3. d	1.2.3
5. b	1.2.4
7. b	1.3.1
9. c	1.4.3

MODULE 29402

Answer	Section Reference
1. d	1.2.3
3. d	1.3.1
5. a	1.4.1
7. d	1.5.3
9. b	1.6.2

MODULE 29403

Answer	Section Reference
1. a	1.1.2
3. c	1.2.3
5. a	1.4.0
7. d	1.4.1
9. a	1.5.1

MODULE 29405

Answer	Section Reference
1. b	1.0.0
3. d	1.1.3
5. d	1.1.2
7. b	2.1.2
9. d	2.2.1

ADDITIONAL RESOURCES

29401 GMAW – Aluminum Plate

Aluminum GMAW Welding Guide (PDF). The Lincoln Electric Company, Cleveland, OH, USA. **www.lincolnelectric.com**.

AWS C5.6-89R, Recommended Practices for Gas Metal Arc Welding. Miami, FL: American Welding Society.

GMAW Welding Guide (PDF). The Lincoln Electric Company, Cleveland, OH, USA. **www.lincolnelectric.com**.

Guide for Aluminum Welding. 2013. Troy, OH: Hobart Brothers Corporation.

OSHA 1910.252, Welding, Cutting and Brazing. Washington, DC: Occupational Safety and Health Administration (OSHA).

OSHA 1926.351, Arc Welding and Cutting. Latest Edition. Washington, DC: Occupational Safety and Health Administration (OSHA).

The Procedure Handbook of Arc Welding. 2000. Cleveland, OH: The James F. Lincoln Arc Welding Foundation.

Welding Aluminum: Theory and Practice. 2002. New York, NY: The Aluminum Association.

Welding Handbook, Volumes 1 and 2. Latest Edition. Welding Science & Technology. Miami, FL: American Welding Society.

29404 GMAW – Aluminum Pipe

Aluminum GMAW Welding Guide (PDF). The Lincoln Electric Company, Cleveland, OH, USA. **www.lincolnelectric.com**.

AWS C5.6-89R, Recommended Practices for Gas Metal Arc Welding. Miami, FL: American Welding Society.

D1.2/D1.2M:2014 Structural Welding Code - Aluminum, American Welding Society.

GMAW Welding Guide (PDF). The Lincoln Electric Company, Cleveland, OH, USA. **www.lincolnelectric.com**.

Guide for Aluminum Welding. 2013. Troy, OH: Hobart Brothers Corporation.

OSHA 1910.252, Welding, Cutting and Brazing. Washington, DC: Occupational Safety and Health Administration (OSHA).

OSHA 1926.351, Arc Welding and Cutting. Latest Edition. Washington, DC: Occupational Safety and Health Administration (OSHA).

The Procedure Handbook of Arc Welding. 2000. Cleveland, OH: The James F. Lincoln Arc Welding Foundation.

Welding Aluminum: Theory and Practice. 2002. New York, NY: The Aluminum Association.

Welding Handbook, Volumes 1 and 2. Latest Edition. Welding Science & Technology. Miami, FL: American Welding Society.

29402 GTAW – Aluminum Plate

AWS D1.2/D1.2M, Structural Welding Code—Aluminum. Miami, FL: American Welding Society.

AWS F4.1, Safe Practices for the Preparation of Containers and Piping for Welding, Cutting, and Allied Processes. Miami, FL: American Welding Society.

Gas Tungsten Arc Welding. Latest Edition. Appleton, WI: Miller Electric Manufacturing Company.

Gas Tungsten Arc Welding. Latest Edition. Cleveland, OH: The James F. Lincoln Arc Welding Foundation.

OSHA 1910.252, Welding, Cutting and Brazing. Washington, DC: Occupational Safety and Health Administration (OSHA).

OSHA 1926.351, Arc Welding and Cutting. Latest Edition. Washington, DC: Occupational Safety and Health Administration.

The Procedure Handbook of Arc Welding. Latest Edition. Cleveland, OH: The James F. Lincoln Arc Welding Foundation.

Welding Aluminum: Theory and Practice. 5th Edition. Arlington, VA: The Aluminum Association, Inc.

Welding Handbook, Volumes 1 and 2. Latest Edition. Welding Science & Technology. Miami, FL: The American Welding Society.

29403 GTAW – Aluminum Pipe

AWS D1.2/D1.2M Structural Welding Code—Aluminum. Miami, FL: American Welding Society.

AWS D10.7, Guide for the Gas Shield Arc Welding of Aluminum and Aluminum Alloy Pipe. Miami, FL: American Welding Society.

AWS F4.1, Safe Practices for the Preparation of Containers and Piping for Welding, Cutting, and Allied Processes. Miami, FL: American Welding Society.

Gas Tungsten Arc Welding. Latest Edition. Appleton, WI: Miller Electric Manufacturing Company.

Gas Tungsten Arc Welding. Latest Edition. Cleveland, OH: The James F. Lincoln Arc Welding Foundation.

Guide for Aluminum Welding. 2013. Troy, OH: Hobart Brothers Corporation.

OSHA 1926.351, Arc Welding and Cutting. Latest Edition. Washington, DC: Occupational Safety and Health Administration.

The Procedure Handbook of Arc Welding. Latest Edition. Cleveland, OH: The James F. Lincoln Arc Welding Foundation.

Welding Aluminum: Theory and Practice. 5th Edition. Arlington, VA: The Aluminum Association, Inc.

Welding Handbook, Volumes 1 and 2. Latest Edition. Welding Science & Technology. Miami, FL: The American Welding Society.

29405 Soldering and Brazing of Copper

A Guide to Brazing and Soldering (PDF). The Harris Products Group, Mason, OH.

Brazing, American Society of Metals.

Brazing Handbook, American Welding Society.

C3.11M/C3.11, Specifications for Torch Soldering, American Welding Society.

C3.4M/C3.4, Specification for Torch Brazing, American Welding Society.

Filler Metal Selection Chart (PDF). The Harris Products Group, Mason, OH.

Gas Welding, Cutting, Brazing & Heating Torches Instruction Manual (PDF). The Harris Products Group, Mason, OH.

Guidelines for Hand Soldering Practices, American Welding Society.

Soldering Handbook, American Welding Society.

GLOSSARY

Acetone: A colorless organic solvent that is volatile and extremely flammable. In reference to the soldering and welding process, it is used as a carrier for acetylene gas in cylinders.

Alloy: Any substance made up of two or more metals.

Anodic coating: An artificial buildup of aluminum oxides on the surface of aluminum alloys that improves corrosion resistance. This coating is many times thicker than a coating of naturally occurring oxides.

Artificial aging: A process following solution heat treatment and other processes that allows the alloy to rest for a prescribed period of time at an elevated temperature.

Axis: The center line running the length of an object. In welding, the term *axis* is most often used in relation to an electrode or a section of pipe.

Brazing: A method of joining metals with a nonferrous (see *nonferrous*) filler metal using heat above 800°F (427°C) but below the melting point of the base metals being joined.

Capillary action: The movement of a liquid along the surface of a solid in a spreading action.

Cast aluminum: Aluminum alloys that are molded into their final form by one of various methods such as sand-casting, die-casting, or pressure die-casting. Cast aluminum contains high levels of silicon to enhance the material's ability to form complicated castings.

Coalescence: The growing together or growth into one body of the materials being joined.

Cup depth: The distance that a tube inserts into a fitting, usually determined by a stop inside the fitting.

Electrode extension: The length of the wire that extends beyond the tip of the welding gun contact tube.

Feather: To grind the start and/or stop areas of a weld to a tapered edge.

Fillet: A rounded internal corner or shoulder of filler metal that often appears at the meeting point of a piece of tubing and a fitting when the joint is soldered or brazed.

Flashback arrestors: A type of valve that prevents a flame from traveling back from the torch tip and into the hoses.

Flux: A chemical substance that prevents oxides from forming on the surface of metals as they are heated for soldering, brazing, or welding.

Ingot: A mass of metal shaped for convenient storage and transport.

Liquation cracks: Liquation is the process of separation by melting, such as in an alloy when one element melts while the others remain solid. Liquation cracks result from localized melting in areas such as the grain boundaries, while adjacent grains remain solid. Both solidification and liquation cracking may be referred to as *hot cracking*.

Natural aging: A process that follows solution heat treatment and other processes that allows the alloy to simply rest at room temperature for a prescribed period of time.

Nonferrous: Relating to a group of metals and metal alloys that contain no iron.

Oxidation: The process by which the oxygen in the air combines with metal, producing tarnish and rust.

Pipe axis: An imaginary line that runs longitudinally (lengthwise) through the center of a pipe.

Potentiometer: A three-terminal resistor with an adjustable center connection; used for variable voltage control.

Remaining passes: A nonstandard term used to describe the fill and cover passes as a group.

Root pass: The initial weld pass made to produce a root bead.

Solder: A fusible alloy used to join metals, with a melting point below 800°F (427°C).

Soldering: A method of joining metals with a nonferrous (see *nonferrous*) filler metal using heat below 800°F (427°C) and below the melting point of the base metals being joined. Also known as *soft soldering* or *sweating*.

Solidification cracks: Cracks that occur as the filler and base metal solidify, when the remaining liquid weld metal is unable to fill the spaces between areas of solidifying weld metal. Both solidification and liquation cracking may be referred to as *hot cracking*.

Solution heat treatment: A process applied to a heat-treatable aluminum alloy that raises the temperature of the alloy to roughly 1,000°F (538°C) for a set period of time, causing the alloy elements to disperse more consistently throughout the metal. The metal is then quenched with water to ensure the dispersed elements remain in place within the metal structure, rather than precipitate out.

Standoff distance: The distance from the gas nozzle or insulating nozzle to the workpiece.

Stickout: The distance from the gas nozzle or insulating nozzle to the end of the electrode.

Truncated: Having the pointed top (vertex) of a conical shape cut off, leaving only a plane section parallel to the base of the cone.

Tungsten: A nonstandard common name for the tungsten electrode used in GTAW; also, the principal metal from which the electrodes are constructed.

Wetting: A process that reduces the surface tension so that molten (liquid) solder flows evenly throughout the joint.

Wrought aluminum: Aluminum alloys that are initially formed into ingots or similar shapes and are then worked into their desired shape through bending, rolling, extrusion, or other mechanical methods.